Pitman Research Notes in Mathematics Series

Submission of proposals for consideration
Suggestions for publication, in the form of outlines and representative samples, are invited by the Editorial Board for assessment. Intending authors should approach one of the main editors or another member of the Editorial Board, citing the relevant AMS subject classifications. Alternatively, outlines may be sent directly to the publisher's offices. Refereeing is by members of the board and other mathematical authorities in the topic concerned, throughout the world.

Preparation of accepted manuscripts
On acceptance of a proposal, the publisher will supply full instructions for the preparation of manuscripts in a form suitable for direct photo-lithographic reproduction. Specially printed grid sheets can be provided and a contribution is offered by the publisher towards the cost of typing. Word processor output, subject to the publisher's approval, is also acceptable.

Illustrations should be prepared by the authors, ready for direct reproduction without further improvement. The use of hand-drawn symbols should be avoided wherever possible, in order to maintain maximum clarity of the text.

The publisher will be pleased to give any guidance necessary during the preparation of a typescript, and will be happy to answer any queries.

Important note
In order to avoid later retyping, intending authors are strongly urged not to begin final preparation of a typescript before receiving the publisher's guidelines. In this way it is hoped to preserve the uniform appearance of the series.

Longman Scientific & Technical
Longman House
Burnt Mill
Harlow, Essex, CM20 2JE
UK
(Telephone (0279) 426721)

Titles in this series. A full list is available on request from the publisher.

100 Optimal control of variational inequalities
V Barbu

101 Partial differential equations and dynamical systems
W E Fitzgibbon III

102 Approximation of Hilbert space operators Volume II
C Apostol, L A Fialkow, D A Herrero and D Voiculescu

103 Nondiscrete induction and iterative processes
V Ptak and F-A Potra

104 Analytic functions - growth aspects
O P Juneja and G P Kapoor

105 Theory of Tikhonov regularization for Fredholm equations of the first kind
C W Groetsch

106 Nonlinear partial differential equations and free boundaries. Volume I
J I Díaz

107 Tight and taut immersions of manifolds
T E Cecil and P J Ryan

108 A layering method for viscous, incompressible L_p flows occupying R^n
A Douglis and E B Fabes

109 Nonlinear partial differential equations and their applications: Collège de France Seminar. Volume VI
H Brezis and J L Lions

110 Finite generalized quadrangles
S E Payne and J A Thas

111 Advances in nonlinear waves. Volume II
L Debnath

112 Topics in several complex variables
E Ramírez de Arellano and D Sundararaman

113 Differential equations, flow invariance and applications
N H Pavel

114 Geometrical combinatorics
F C Holroyd and R J Wilson

115 Generators of strongly continuous semigroups
J A van Casteren

116 Growth of algebras and Gelfand–Kirillov dimension
G R Krause and T H Lenagan

117 Theory of bases and cones
P K Kamthan and M Gupta

118 Linear groups and permutations
A R Camina and E A Whelan

119 General Wiener–Hopf factorization methods
F-O Speck

120 Free boundary problems: applications and theory. Volume III
A Bossavit, A Damlamian and M Fremond

121 Free boundary problems: applications and theory. Volume IV
A Bossavit, A Damlamian and M Fremond

122 Nonlinear partial differential equations and their applications: Collège de France Seminar. Volume VII
H Brezis and J L Lions

123 Geometric methods in operator algebras
H Araki and E G Effros

124 Infinite dimensional analysis–stochastic processes
S Albeverio

125 Ennio de Giorgi Colloquium
P Krée

126 Almost-periodic functions in abstract spaces
S Zaidman

127 Nonlinear variational problems
A Marino, L Modica, S Spagnolo and M Degliovanni

128 Second-order systems of partial differential equations in the plane
L K Hua, W Lin and C-Q Wu

129 Asymptotics of high-order ordinary differential equations
R B Paris and A D Wood

130 Stochastic differential equations
R Wu

131 Differential geometry
L A Cordero

132 Nonlinear differential equations
J K Hale and P Martinez-Amores

133 Approximation theory and applications
S P Singh

134 Near-rings and their links with groups
J D P Meldrum

135 Estimating eigenvalues with *a posteriori/a priori* inequalities
J R Kuttler and V G Sigillito

136 Regular semigroups as extensions
F J Pastijn and M Petrich

137 Representations of rank one Lie groups
D H Collingwood

138 Fractional calculus
G F Roach and A C McBride

139 Hamilton's principle in continuum mechanics
A Bedford

140 Numerical analysis
D F Griffiths and G A Watson

141 Semigroups, theory and applications. Volume I
H Brezis, M G Crandall and F Kappel

142 Distribution theorems of L-functions
D Joyner

143 Recent developments in structured continua
D De Kee and P Kaloni

144 Functional analysis and two-point differential operators
J Locker

145 Numerical methods for partial differential equations
S I Hariharan and T H Moulden

146 Completely bounded maps and dilations
V I Paulsen

147 Harmonic analysis on the Heisenberg nilpotent Lie group
W Schempp

148 Contributions to modern calculus of variations
L Cesari

149 Nonlinear parabolic equations: qualitative properties of solutions
L Boccardo and A Tesei

150 From local times to global geometry, control and physics
K D Elworthy

151 A stochastic maximum principle for optimal control of diffusions
U G Haussmann

152 Semigroups, theory and applications. Volume II
H Brezis, M G Crandall and F Kappel

153 A general theory of integration in function spaces
P Muldowney

154 Oakland Conference on partial differential equations and applied mathematics
L R Bragg and J W Dettman

155 Contributions to nonlinear partial differential equations. Volume II
J I Díaz and P L Lions

156 Semigroups of linear operators: an introduction
A C McBride

157 Ordinary and partial differential equations
B D Sleeman and R J Jarvis

158 Hyperbolic equations
F Colombini and M K V Murthy

159 Linear topologies on a ring: an overview
J S Golan

160 Dynamical systems and bifurcation theory
M I Camacho, M J Pacifico and F Takens

161 Branched coverings and algebraic functions
M Namba

162 Perturbation bounds for matrix eigenvalues
R Bhatia

163 Defect minimization in operator equations: theory and applications
R Reemtsen

164 Multidimensional Brownian excursions and potential theory
K Burdzy

165 Viscosity solutions and optimal control
R J Elliott

166 Nonlinear partial differential equations and their applications: Collège de France Seminar. Volume VIII
H Brezis and J L Lions

167 Theory and applications of inverse problems
H Haario

168 Energy stability and convection
G P Galdi and B Straughan

169 Additive groups of rings. Volume II
S Feigelstock

170 Numerical analysis 1987
D F Griffiths and G A Watson

171 Surveys of some recent results in operator theory. Volume I
J B Conway and B B Morrel

172 Amenable Banach algebras
J-P Pier

173 Pseudo-orbits of contact forms
A Bahri

174 Poisson algebras and Poisson manifolds
K H Bhaskara and K Viswanath

175 Maximum principles and eigenvalue problems in partial differential equations
P W Schaefer

176 Mathematical analysis of nonlinear, dynamic processes
K U Grusa

177 Cordes' two-parameter spectral representation theory
D F McGhee and R H Picard

178 Equivariant K-theory for proper actions
N C Phillips

179 Elliptic operators, topology and asymptotic methods
J Roe

180 Nonlinear evolution equations
J K Engelbrecht, V E Fridman and E N Pelinovski

181 Nonlinear partial differential equations and their applications: Collège de France Seminar. Volume IX
H Brezis and J L Lions

182 Critical points at infinity in some variational problems
A Bahri

183 Recent developments in hyperbolic equations
L Cattabriga, F Colombini, M K V Murthy and S Spagnolo

184 Optimization and identification of systems governed by evolution equations on Banach space
N U Ahmed

185 Free boundary problems: theory and applications. Volume I
K H Hoffmann and J Sprekels

186 Free boundary problems: theory and applications. Volume II
K H Hoffmann and J Sprekels

187 An introduction to intersection homology theory
F Kirwan

188 Derivatives, nuclei and dimensions on the frame of torsion theories
J S Golan and H Simmons

189 Theory of reproducing kernels and its applications
S Saitoh

190 Volterra integrodifferential equations in Banach spaces and applications
G Da Prato and M Iannelli

191 Nest algebras
K R Davidson

192 Surveys of some recent results in operator theory. Volume II
J B Conway and B B Morrel

193 Nonlinear variational problems. Volume II
A Marino and M K V Murthy

194 Stochastic processes with multidimensional parameter
M E Dozzi

195 Prestressed bodies
D Iesan

196 Hilbert space approach to some classical transforms
R H Picard

197 Stochastic calculus in application
J R Norris

198 Radical theory
B J Gardner

199 The C^*-algebras of a class of solvable Lie groups
X Wang

200 Stochastic analysis, path integration and dynamics
K D Elworthy and J C Zambrini

201 Riemannian geometry and holonomy groups
S Salamon
202 Strong asymptotics for extremal errors and polynomials associated with Erdős type weights
D S Lubinsky
203 Optimal control of diffusion processes
V S Borkar
204 Rings, modules and radicals
B J Gardner
205 Two-parameter eigenvalue problems in ordinary differential equations
M Faierman
206 Distributions and analytic functions
R D Carmichael and D Mitrovic
207 Semicontinuity, relaxation and integral representation in the calculus of variations
G Buttazzo
208 Recent advances in nonlinear elliptic and parabolic problems
P Bénilan, M Chipot, L Evans and M Pierre
209 Model completions, ring representations and the topology of the Pierce sheaf
A Carson
210 Retarded dynamical systems
G Stepan
211 Function spaces, differential operators and nonlinear analysis
L Paivarinta
212 Analytic function theory of one complex variable
C C Yang, Y Komatu and K Niino
213 Elements of stability of visco-elastic fluids
J Dunwoody
214 Jordan decomposition of generalized vector measures
K D Schmidt
215 A mathematical analysis of bending of plates with transverse shear deformation
C Constanda
216 Ordinary and partial differential equations. Volume II
B D Sleeman and R J Jarvis
217 Hilbert modules over function algebras
R G Douglas and V I Paulsen
218 Graph colourings
R Wilson and R Nelson
219 Hardy-type inequalities
A Kufner and B Opic
220 Nonlinear partial differential equations and their applications: Collège de France Seminar. Volume X
H Brezis and J L Lions
221 Workshop on dynamical systems
E Shiels and Z Coelho
222 Geometry and analysis in nonlinear dynamics
H W Broer and F Takens
223 Fluid dynamical aspects of combustion theory
M Onofri and A Tesei
224 Approximation of Hilbert space operators. Volume I. 2nd edition
D Herrero
225 Operator theory: proceedings of the 1988 GPOTS–Wabash conference
J B Conway and B B Morrel
226 Local cohomology and localization
J L Bueso Montero, B Torrecillas Jover and A Verschoren
227 Nonlinear waves and dissipative effects
D Fusco and A Jeffrey
228 Numerical analysis 1989
D F Griffiths and G A Watson
229 Recent developments in structured continua. Volume II
D De Kee and P Kaloni
230 Boolean methods in interpolation and approximation
F J Delvos and W Schempp
231 Further advances in twistor theory. Volume I
L J Mason and L P Hughston
232 Further advances in twistor theory. Volume II
L J Mason and L P Hughston
233 Geometry in the neighborhood of invariant manifolds of maps and flows and linearization
U Kirchgraber and K Palmer
234 Quantales and their applications
K I Rosenthal
235 Integral equations and inverse problems
V Petkov and R Lazarov
236 Pseudo-differential operators
S R Simanca
237 A functional analytic approach to statistical experiments
I M Bomze
238 Quantum mechanics, algebras and distributions
D Dubin and M Hennings
239 Hamilton flows and evolution semigroups
J Gzyl
240 Topics in controlled Markov chains
V S Borkar
241 Invariant manifold theory for hydrodynamic transition
S Sritharan
242 Lectures on the spectrum of $L^2(\Gamma\backslash G)$
F L Williams
243 Progress in variational methods in Hamiltonian systems and elliptic equations
M Girardi, M Matzeu and F Pacella
244 Optimization and nonlinear analysis
A Ioffe, M Marcus and S Reich
245 Inverse problems and imaging
G F Roach
246 Semigroup theory with applications to systems and control
N U Ahmed
247 Periodic-parabolic boundary value problems and positivity
P Hess
248 Distributions and pseudo-differential operators
S Zaidman
249 Progress in partial differential equations: the Metz surveys
M Chipot and J Saint Jean Paulin
250 Differential equations and control theory
V Barbu

251 Stability of stochastic differential equations with respect to semimartingales
X Mao
252 Fixed point theory and applications
J Baillon and M Théra
253 Nonlinear hyperbolic equations and field theory
M K V Murthy and S Spagnolo
254 Ordinary and partial differential equations. Volume III
B D Sleeman and R J Jarvis
255 Harmonic maps into homogeneous spaces
M Black
256 Boundary value and initial value problems in complex analysis: studies in complex analysis and its applications to PDEs 1
R Kühnau and W Tutschke
257 Geometric function theory and applications of complex analysis in mechanics: studies in complex analysis and its applications to PDEs 2
R Kühnau and W Tutschke
258 The development of statistics: recent contributions from China
X R Chen, K T Fang and C C Yang
259 Multiplication of distributions and applications to partial differential equations
M Oberguggenberger
260 Numerical analysis 1991
D F Griffiths and G A Watson
261 Schur's algorithm and several applications
M Bakonyi and T Constantinescu
262 Partial differential equations with complex analysis
H Begehr and A Jeffrey
263 Partial differential equations with real analysis
H Begehr and A Jeffrey
264 Solvability and bifurcations of nonlinear equations
P Drábek
265 Orientational averaging in mechanics of solids
A Lagzdins, V Tamuzs, G Teters and A Kregers
266 Progress in partial differential equations: elliptic and parabolic problems
C Bandle, J Bemelmans, M Chipot, M Grüter and J Saint Jean Paulin
267 Progress in partial differential equations: calculus of variations, applications
C Bandle, J Bemelmans, M Chipot, M Grüter and J Saint Jean Paulin
268 Stochastic partial differential equations and applications
G Da Prato and L Tubaro
269 Partial differential equations and related subjects
M Miranda
270 Operator algebras and topology
W B Arveson, A S Mishchenko, M Putinar, M A Rieffel and S Stratila
271 Operator algebras and operator theory
W B Arveson, A S Mishchenko, M Putinar, M A Rieffel and S Stratila
272 Ordinary and delay differential equations
J Wiener and J K Hale
273 Partial differential equations
J Wiener and J K Hale
274 Mathematical topics in fluid mechanics
J F Rodrigues and A Sequeira
275 Green functions for second order parabolic integro-differential problems
M G Garroni and J F Menaldi
276 Riemann waves and their applications
M W Kalinowski
277 Banach C(K)-modules and operators preserving disjointness
Y A Abramovich, E L Arenson and A K Kitover
278 Limit algebras: an introduction to subalgebras of C*-algebras
S C Power
279 Abstract evolution equations, periodic problems and applications
D Daners and P Koch Medina
280 Emerging applications in free boundary problems
J M Chadam and H Rasmussen
281 Free boundary problems involving solids
J M Chadam and H Rasmussen
282 Free boundary problems in fluid flow with applications
J M Chadam and H Rasmussen
283 Asymptotic problems in probability theory: stochastic models and diffusions on fractals
K D Elworthy and N Ikeda
284 Asymptotic problems in probability theory: Wiener functionals and asymptotics
K D Elworthy and N Ikeda
285 Dynamical systems
R Bamon, R Labarca, J Lewowicz and J Palis
286 Models of hysteresis
A Visintin
287 Moments in probability and approximation theory
G Anastassiou
288 Mathematical aspects of penetrative convection
B Straughan
289 Ordinary and partial differential equations. Vol ume IV
B D Sleeman and R J Jarvis
290 *K*-theory for real *C**-algebras
H Schröder
291 Recent developments in theoretical fluid mechanics
G P Galdi and J Necas
292 Propagation of a curved shock and nonlinear ray theory
P Prasad
293 Non-classical elastic solids
M Ciarletta and D Ieșan
294 Multigrid methods
J Bramble
295 Entropy and partial differential equations
W A Day
296 Progress in partial differential equations: the Metz surveys 2
M Chipot
297 Nonstandard methods in the calculus of variations
C Tuckey
298 Barrelledness, Baire-like- and (LF)-spaces
M Kunzinger
299 Nonlinear partial differential equations and their applications. Collège de France Seminar. Volume XI
H Brezis and J L Lions
300 Introduction to operator theory
T Yoshino

301 Generalized fractional calculus and applications
V Kiryakova
302 Nonlinear partial differential equations and their applications. Collège de France Seminar Volume XII
H Brezis and J L Lions
303 Numerical analysis 1993
D F Griffiths and G A Watson
304 Topics in abstract differential equations
S Zaidman
305 Complex analysis and its applications
C C Yang, G C Wen, K Y Li and Y M Chiang
306 Computational methods for fluid-structure interaction
J M Crolet and R Ohayon
307 Random geometrically graph directed self-similar multifractals
L Olsen
308 Progress in theoretical and computational fluid mechanics
G P Galdi, J Málek and J Necas
309 Variational methods in Lorentzian geometry
A Masiello
310 Stochastic analysis on infinite dimensional spaces
H Kunita and H-H Kuo
311 Representations of Lie groups and quantum groups
V Baldoni and M Picardello
312 Common zeros of polynomials in several variables and higher dimensional quadrature
Y Xu
313 Extending modules
N V Dung, D van Huynh, P F Smith and R Wisbauer
314 Progress in partial differential equations: the Metz surveys 3
M Chipot, J Saint Jean Paulin and I Shafrir
315 Refined large deviation limit theorems
V Vinogradov
316 Topological vector spaces, algebras and related areas
A Lau and I Tweddle
317 Integral methods in science and engineering
C Constanda
318 A method for computing unsteady flows in porous media
R Raghavan and E Ozkan

R Raghavan

Phillips Petroleum Company, USA

and

E Ozkan

Istanbul Technical University, Turkey

A method for computing unsteady flows in porous media

CRC Press
Taylor & Francis Group
Boca Raton London New York

CRC Press is an imprint of the
Taylor & Francis Group, an **informa** business
A CHAPMAN & HALL BOOK

CRC Press
Taylor & Francis Group
6000 Broken Sound Parkway NW, Suite 300
Boca Raton, FL 33487-2742

First issued in hardback 2017

ISBN 13: 978-1-138-40409-0 (hbk)
ISBN 13: 978-0-582-07811-6 (pbk)

Visit the Taylor & Francis Web site at
http://www.taylorandfrancis.com

and the CRC Press Web site at
http://www.crcpress.com

British Library Cataloguing in Publication Data

A catalogue record for this book is
available from the British Library

Library of Congress Cataloging-in-Publication Data

Raghavan, R.
A method for computing unsteady flows in porous media / R. Raghavan & E. Ozkan.
p. cm. -- (Pitman research notes in mathematics series, ISSN 0269-3674)
1. Unsteady flow (Fluid dynamics) 2. Differential equations, Partial--Numerical solutions. 3. Porous materials. I. Ozkan, E. II. Title. III. Series.
QA911.R26 1992
532'.052--dc20 92-4745
CIP

Contents

Preface

We are stimulated to write this survey for at least two reasons. First, much of the information on flow through porous media appears in disparate journals and there is little or no interdisciplinary communication. Second, a number of researchers, particularly in petroleum engineering, have spent considerable effort over the past thirty years developing computationally-efficient algorithms to describe flow through porous media with a view to evaluating the properties of subterranean reservoirs. A compilation of recent advances should be of particular interest to those involved in the study of partial differential equations.

The issues we address arise in a number of scientific and engineering disciplines. Much of the impetus for the study of flow through porous media, however, derives as a consequence of flow in geologic media—the work of hydrologists, soil physicists, and agricultural engineers who work at shallow depths and petroleum engineers who, lured by the need to produce oil and gas more efficiently, operate at deeper levels. In chemical engineering, packed bed reactors are used to facilitate various reactions. Environmental concerns have also led to an increased interest in flow through porous media.

Briefly, these notes may be thought of as a survey of implicit evolution equations that arise in classical models of diffusion or convection in composite materials. Restricting our attention to linear systems, the mathematical description of the diffusion process in a composite medium consisting of two components leads to the following pair of equations:

$$c_1 u_t - a_1 \nabla^2 u + \frac{1}{\epsilon}(u - v) = 0, \tag{1}$$

and

$$c_2 v_t - a_2 \nabla^2 v + \frac{1}{\epsilon}(v - u) = 0. \tag{2}$$

Here, the dependent variables $u[(\mathbf{x}), t]$ and $v[(\mathbf{x}); t]$ denote temperatures, densities, or pressures in the respective components at $t > 0$, averaged at the point $\mathbf{x} \in R_n$ over a neighborhood containing both components. c_1 and c_2 are monotone functions that represent fluid storage-capacity or heat content, a_1 and a_2 are non-negative functions that correspond to permeability or conductivity, and ϵ models the exchange rate between the two components. Our principal focus is on cases wherein $a_2/c_2 >> a_1/c_1$ which occur when $c_2 \approx 0$—the fissured-medium equation, or the Barenblatt problem. The Barenblatt problem does not specify details regarding the exchange of fluid between the two media. We also consider models that examine fluid exchange between the matrix blocks and the fissures in greater detail.

In most geologic settings, it is not always possible to deal with the porous medium as if it were a single layer (even in the contexts noted above). Stratification cannot be ignored. The extension of (1) and (2) to incorporate the layered nature of the porous medium is presented. When developing mathematical models to many of the settings noted above, it is essential to incorporate the existence of extraction points (wellbores). In many realistic situations the geometrical shapes of the extraction points can be quite complex. Thus, although we consider linear systems of the form noted in (1) and (2), procedures to obtain solutions when boundary conditions are incorporated can become rather formidable.

Simply stated, these notes provide a basis for the development of algorithms for the study of unsteady flow in saturated porous media. These algorithms provide for the examination of three-dimensional problems and complicated boundary conditions that are a natural consequence of flow in geologic media. Chapter I presents an overview of flow through porous media, previews notation and addresses a few basic issues for ease of understanding. Chapter II considers the fundamental solution for flow in porous media in Cartesian and cylindrical reference frames subject to Dirichlet, Neumann, and mixed boundary-conditions. Chapter III demonstrates how one arrives at the expressions for pressure distributions in porous media that account for the extraction points. For illustrative purposes, we assume that the extraction points may be represented by lines and discs (circles or rectangles). Chapter IV, the crux of this work, is intended to aid those who are interested in developing algorithms for computing pressure distributions. Solutions in Chapter III are reformulated with a view to aid computations. The observations noted here are based on our experience, and many of the solutions presented here have been computed with the numerical algorithm of Stehfest (Communications of the ACM, January 1970, page 47). Asymptotic forms of various solutions are also given here. Chapter V is a natural progression of the solutions in Chapter III to more complicated visualizations of flow in porous media.

It is our pleasure to thank four individuals who were indispensable to this work. This manuscript was typed in its entirety by Ms. Jan Want and we are grateful for her patience in going through the myriad revisions of this work. Ms. Kathleen Henzel of the Tulsa Public Schools helped us with editing the manuscript. Mr. Chih-Cheng Chen of Halliburton Energy Services verified many of the developments given here and provided us with valuable comments. We thank Ms. Laura Passiglia for drafting the figures. In spite of all the assistance we have received, we are responsible for the blemishes that remain. We thank the Society of Petroleum Engineers (SPE) for permission to reproduce Figure 5.1 in Chapter V of this book.

R.R.
E.O.

Nomenclature

B:	boundary
C:	ability of wellbore to store fluid per unit change in pressure
c:	compressibility (fluid)
C_D:	wellbore storage coefficient
C_{fD}:	dimensionless conductivity of fracture
C_{hD}:	dimensionless conductivity of horizontal well
c_m:	compressibility (porous medium)
c_t:	compressibility (total)
$ch(x)$:	hyperbolic cosine
D:	space-time domain, normalized (subscript)
$-Ei(-x)$:	exponential integral function
$\mathrm{erf}(x)$:	error function
$\mathrm{erfc}(x)$:	complementary error function
f:	source density, fanning friction factor, fissure system (subscript)
$f(M;t)$:	mass rate
$f(s)$:	function to incorporate fissured-medium characteristics
$\overline{f}(s)$:	Laplace transform of $F(t)$
f_t:	friction factor
h:	thickness
h_f:	length of a line source
$I_{m+1/2}(x)$:	modified spherical Bessel function
$I_\nu(x)$:	modified Bessel function
k:	permeability
k_f:	fracture permeability
k_i:	permeability in the i-direction
$Ki_n(x)$:	repeated integral of $K_0(x)$
$K_{m+1/2}(x)$:	modified spherical Bessel function
$K_\nu(x)$:	modified Bessel function
L:	diffusion operator
$\mathcal{L}$:	Laplace transform operator
l:	characteristic length
L_f:	half-length of a vertical fracture
L_h:	length of a horizontal well
$\mathbf{L}_\nu(x)$:	modified Struve function
$\overline{L}$:	symbolic Laplace transform of L
L^*:	formal adjoint of L
m:	blocks of fissured porous media (subscript)
M:	point, position vector

n:	outward normal to the boundary
p:	pressure
p_f:	fissure-system pressure
$p_h(x)$:	pressure in the horizontal well
p_i:	initial value of p
p_m:	matrix-system pressure
$\mathrm{P}_m(x)$:	Legendre polynomial
$P_m(x)$:	Legendre polynomial
$\mathrm{P}_m^k(x)$:	associated Legendre function
p_{wf}:	wellbore pressure
q:	withdrawal rate from porous medium
$Q(t)$:	volumetric rate
$q_{hc}(x)$:	flow rate at any point in the wellbore
Q_{pi}:	point-source density (instantaneous)
Q_{pc}:	point-source density (continuous)
q_{sf}:	influx into the wellbore
$\widetilde{q}$:	volumetric production rate per unit volume
R:	subdomain, distance between points P and P'
r:	distance, r-coordinate
$\mathrm{R}e$:	Reynolds number
r_e:	distance to the boundary
$\mathrm{R}e_t$:	Reynolds number based on production rate
R_n:	n-dimensional Euclidean space
S:	normalized pressure drop in the skin region
s:	Laplace transform variable
$sh(x)$:	hyperbolic sine
T:	time interval; temperature
t:	time variable
$th(x)$:	hyperbolic tangent
u:	equals s or $sf(s)$
V:	volume, bulk volume
V_f:	fluid volume
V_j:	ratio of the volume of system j to total volume
$v(M;t)$:	fluid velocity
V_p:	pore volume
V_w:	wellbore volume
w_f:	width of a vertical fracture
x:	distance, x-coordinate
x_e:	distance to the boundary
y:	distance, y-coordinate
y_e:	distance to the boundary
z:	distance, z-coordinate

z_e: distance to the boundary
α: angle
Γ: boundary
$\Gamma(x)$: Gamma function
γ: fundamental solution, Euler's constant (0.5772...)
$\delta(M;t)$: Dirac-delta function
$1/\epsilon$: degree of fissuring
η: diffusivity
θ: angle, coordinate
λ: characteristic constant of the fissured system
μ: viscosity
ν: angle, order (subscript)
ρ: distance, spherical coordinate
ρ_f: radius of a disc source
$\rho(M;t)$: density
ϕ: porosity, angle, coordinate
$\phi(M;t)$: test function
φ: angle
ψ: angle
Ω: space domain
ω: characteristic constant of the fissured system

I. General Theory

Our primary interest is to consider three-dimensional flow in porous media, including fissured and layered media. We propose to provide for extraction or injection of fluids via points that are complicated in geometry for a variety of conditions on the boundaries of the porous solid. Here, we preview our notation and address a few basic issues for ease of understanding.

Our notation is standard. The symbols x, y, and z denote the space variables in Cartesian coordinates and t denotes the time variable. Ω is a bounded domain in space and Γ is the boundary of Ω. The position of a point in Ω is defined by the position vector $\mathbf{M} = (x, y, z)$. T denotes the time interval $T; \{t|0 < t < \infty\}$. The space-time domain $D = \Omega \times T$ is the product of the region Ω and the time interval T. An inhomogeneous porous medium occupies Ω and a fluid (liquid or gas) flows through the porous medium. The porosity, $\phi(\mathbf{M}; t)$, of the medium (defined as the void volume, V_p, per bulk volume, V), and the density, $\rho(\mathbf{M}; t)$, of the fluid are both scalar functions of $\mathbf{M}$ and t, for $\mathbf{M}$ and $t \in D$. Sources or sinks in D may supply or extract fluid. In the following development, we use the term *source* to denote both the supply and the extraction of fluid with the understanding that the strength of a sink is negative. We also drop the distinguishing notation for vectors.

1.1. Equations for flow through porous solids

Let R be an arbitrary subdomain of Ω with boundary B, and n be the outward normal to B. Let $v(M; t)$ denote the velocity of the fluid, and $f(M; t)$ denote the local rate (mass per unit volume per unit time) at which mass is extracted from R. The conservation of mass principle yields

$$\frac{\partial}{\partial t} \int_R \rho\phi dM = - \int_B \rho v \cdot n d\Gamma - \int_R f(M; t) dM. \tag{1}$$

A local version of (1) is obtained by applying the Divergence theorem to the surface integral in (1) and is given by

$$\frac{\partial}{\partial t}(\rho\phi) = -\nabla \cdot (\rho v) - f; \ (M; t) \in D. \tag{2}$$

This development assumes that the functions involved in (2) are continuous and hence excludes the existence of sources in D that are impulsive in time and/or concentrated in space. To incorporate impulsive and concentrated sources, we provide a distributional interpretation of (2) in §1.2.

We now assume that Darcy's Law [1] holds. We restrict our attention to Newtonian fluids and isothermal conditions. Ignoring gravity, the velocity of the fluid is given by

$$v = -\frac{k}{\mu} \nabla p, \tag{3}$$

where k is the permeability of the medium, μ is the viscosity of the fluid, and p is the pressure. For the present, we assume that $k = k(M)$. Later we will restrict our attention to an anisotropic solid in which the permeability of the porous medium is independent of position, and the coordinate axes are coincident with the principal axes of permeability, see Childs [2].

The isothermal compressibility of the fluid, c, and the compressibility of the porous medium, c_m, are defined, respectively, by

$$c = \left(\frac{1}{\rho}\frac{\partial \rho}{\partial p}\right)_T = -\left(\frac{1}{V_f}\frac{\partial V_f}{\partial p}\right)_T \tag{4}$$

and

$$c_m = \left(\frac{1}{V_p}\frac{\partial V_p}{\partial p}\right)_T \approx \left(\frac{1}{\phi}\frac{\partial \phi}{\partial p}\right)_T, \tag{5}$$

where V_f is the fluid volume and V_p is the pore volume. The total compressibility of the system, then, is $c_t = c + c_m$. A comprehensive survey of pore-volume compressibility is given by Scorer and Miller [3].

Insertion of the expressions on the right-hand sides of (3)—(5) for the appropriate expressions in (2) yields the following diffusion equation in terms of pressure, p:

$$\phi c_t \frac{\partial p}{\partial t} = \nabla \cdot \left(\frac{k}{\mu}\nabla p\right) + \frac{k}{\mu} c \left(\nabla p\right)^2 - \tilde{q}, \tag{6}$$

where $\tilde{q} = f/\rho$ is the volumetric, production rate per unit volume. Note that in (6), ϕ and c_t are functions of pressure. The effects of the variations of ϕ and c_t with pressure can be examined along the lines outlined by Raghavan, Scorer, and Miller [4]. Most often this variation is ignored and ϕ and c_t are assumed to be constants. For the purposes of this discussion, we will assume that ϕ is a constant and $c_t = c$. If we assume that the pressure gradients are small and that the compressibility of the fluid and the viscosity of the fluid are constants, then (6) leads to the linear, diffusion equation

$$\nabla \cdot \left(\frac{k}{\mu}\nabla p\right) - \phi c \frac{\partial p}{\partial t} - \tilde{q} = 0. \tag{7}$$

Remark 1: *In the context of heat conduction, p corresponds to temperature, k/μ corresponds to thermal conductivity, and ϕc corresponds to the heat content of the solid.*

Remark 2: *If density and viscosity are functions of pressure, then an equation similar to (7) can be derived from (2) via the Kirchhoff [5] transformation*

$$m(p) = \int_0^p \frac{\rho}{\mu} dp'. \tag{8}$$

This transformation yields a diffusion equation in terms of $m(p)$ and eliminates the need to assume that $c(\nabla p)^2 \approx 0$. In passing, we note that for flow in simple geometrical *systems (rectilinear, cylindrical or spherical), the transformation suggested by Cole [6] and Hopf [7] renders the assumption* $c(\nabla p)^2 \approx 0$ *unnecessary.*

Remark 3: *For a homogeneous and isotropic solid in which permeability is independent of position, (7) becomes*

$$\eta\nabla^2 p - \frac{\partial p}{\partial t} - \frac{\tilde{q}}{\phi c} = 0. \tag{9}$$

Here, $\eta = k/(\phi c\mu)$ *is the "diffusivity" of the porous solid. For solids in which the principal axes of permeability coincide with the coordinate axes, (9) will also describe the diffusion process in the transformed coordinates* $i' = i\sqrt{k/k_i}$ *for* $i = x, y,$ *or* z *where* k *may be chosen arbitrarily and* k_i *represents the permeability in the* i*-direction (note that if* k *is chosen arbitrarily, then* $\tilde{q}(M';t)$ *is not necessarily equal to* $\tilde{q}(M;t)$*; if, however,* k *is chosen to be* $(k_x k_y k_z)^{\frac{1}{3}}$*, then* $dM' = dM$ *and* $\tilde{q}(M';t) = \tilde{q}(M;t)$*). If the solid is infinite in extent or is bounded by planes perpendicular to the principal axes of permeability, this transformation reduces the problem of flow in an anisotropic solid to that of flow in an isotropic solid. In other cases, the bounding surfaces are usually distorted.*

Remark 4: *If* $Q(t)$ *is the volumetric rate at which sources supply fluid in* R*, then* $Q(t)/(\phi c)$ *is the* strength *of the sources in* R*. Because*

$$\frac{Q(t)}{\phi c} = \frac{1}{\phi c}\int_R \tilde{q}(M;t)dM, \tag{10}$$

$\tilde{q}(M;t)/(\phi c)$ *in (10) represents the density of the sources of strength,* $Q(t)/(\phi c)$*. Note that if* V *denotes the volume corresponding to* R*, and if* $P = \int_R dpdM$ *denotes the change in pressure in* R *from time* t *to time* $t + dt$*, then* $P = Q/(\phi c)/V$*.*

For generality and to be more concise, (9) can be expressed in terms of the normalized quantities

$$i_D = i/\ell, \tag{11}$$

where $i = x, y,$ or z, and

$$t_D = \eta t/\ell^2 \tag{12}$$

as follows

$$\nabla_D^2 p - \frac{\partial p}{\partial t_D} - \frac{\tilde{q}_D}{\phi c} = 0; \quad (M_D; t_D) \in D_D. \tag{13}$$

Here, ℓ represents the characteristic length of the system, ∇_D^2 is the Laplacian operator in i_D, D_D is the space-time domain in terms of normalized quantities, and $\tilde{q}_D/(\phi c)$ is the source density in D_D. It is also convenient to define the diffusion operator as

$$L = \nabla_D^2 - \frac{\partial}{\partial t_D} \tag{14}$$

and write (13) as

$$Lp = \frac{\tilde{q}_D}{\phi c}; \ (M_D; t_D) \in D_D. \tag{15}$$

The developments in the later parts of this work are in terms of the Laplace transformation. Application of the Laplace transformation to (15) yields

$$\overline{L}\overline{p}(s) = \frac{\overline{\tilde{q}_D}}{\phi c} - p_i; \quad M_D \in \Omega_D, \tag{16}$$

where

$$\overline{f}(s) = \mathcal{L}\{f(t_D)\} = \int_0^\infty e^{-st_D} f(t_D) dt_D. \tag{17}$$

In (16), $p_i = \lim_{t_D \to 0+} p\,(M_D; t_D)$ represents the initial value of p, and $\overline{L}$ represents the symbolic Laplace transform of L given by

$$\overline{L} = \nabla_D^2 - s. \tag{18}$$

In passing, we note that the elliptic operator, $\overline{L}$, is a self-adjoint operator, whereas the parabolic operator, L, is not a self-adjoint operator.

1.2. Diffusion with impulsive and concentrated sources. Distributions

Here, we develop the framework to consider extraction of fluid from porous media through sources that are nearly impulsive and almost localized. For notational simplicity, we will assume that the porous medium has unit properties.

Let us consider the diffusion equation

$$Lp(M;t) = f(M;t); \quad (M;t) \in D. \tag{1}$$

In the development in §1.1, the nonhomogeneous term, f, of the diffusion equation was required to be a continuous function and p was a sufficiently differentiable function that satisfied (1) pointwise on D. These requirements are met when f corresponds to the density of a continuous and distributed source. For our purposes, however, we need to interpret f as the density of an impulsive and/or a concentrated source. This is readily accomplished by giving meaning to (1) in terms of the Theory of Distributions. The basis for the distributional interpretation of (1) may be found in standard developments of the theory of distributions; see Schwartz [8], Zemanian [9], and Stakgold [10]. For continuity, we present the following definitions (from Stakgold [10]).

Definition I: *An infinitely differentiable function, $\phi(M)$, on R_n with compact support is called a test function on R_n where R_n represents n-dimensional Euclidean space. The space of all test functions on R_n is denoted by $C_0^\infty(R_n)$.*

Definition II: *An n-dimensional distribution, f, is defined by the rule*

$$f = \langle f, \phi \rangle = \int_{R_n} f(M)\phi(M)dM, \tag{2}$$

where $f(M)$ is a function in R_n that is locally integrable and $\phi(M)$ is a test function belonging to $C_0^\infty(R_n)$.

Definition III: *A distribution f is said to be regular if it can be defined through the rule given by (2) with $f(M)$ locally integrable. All other distributions are said to be singular. Given a singular distribution, f, we can assign to it a generalized function, $f(M)$, and still use (2) symbolically.*

Definition IV: *The distributions f_1 and f_2 are said to be equal in the open set Ω if $\langle f_1 - f_2, \phi \rangle = 0$ for every test function, $\phi(M)$, with support in Ω.*

For a given distribution f, we can now interpret (1) as a differential equation in a distributional sense and require that the distributions Lp and f be equal in D. Here, p is a distributional solution of (1) if $\langle p, L^*\phi \rangle = \langle f, \phi \rangle$ for every test function $\phi(M;t)$ in $C_0^\infty(D)$ where L^* is the formal adjoint of L. If f is a distribution generated by a distributed source density as in §1.1 (that is, f is a continuous function), then the sufficiently differentiable function p that satisfies (1) pointwise in D is a classical solution of (1). In the context of the distributional interpretation, (1) still makes sense even if f is a singular distribution. We shall now use the distributional interpretation of (1) to define f as the symbolic density of an impulsive and concentrated source.

Let f be the singular, Dirac delta distribution; therefore

$$Lp(M, \widetilde{M}; t) = \delta(M, \widetilde{M}); \quad (M, \widetilde{M}; t) \in D, \tag{3}$$

is a diffusion equation in terms of distributions and the distribution p is a solution of (3) if

$$\langle p, L^*\phi \rangle = \langle \delta, \phi \rangle = \int_T \int_\Omega \delta(M, \widetilde{M}) \phi(M;t) dM dt = \int_T \phi(\widetilde{M};t) dt. \tag{4}$$

Because $\int_\Omega \delta(M, \widetilde{M}) dM = 1$ for $\widetilde{M}$ in Ω, we can visualize $\delta(M, \widetilde{M})$ as the *symbolic source density for a concentrated source (a point source) of unit strength* at $\widetilde{M}$. If the source strength is different from unity, then we define $f = [\widetilde{q}(\widetilde{M};t)/(\phi c)]\delta(M, \widetilde{M})$ as the *source density* so that

$$\int_\Omega f(M, \widetilde{M}; t) dM = \int_\Omega \frac{\widetilde{q}(\widetilde{M};t)}{\phi c} \delta(M, \widetilde{M}) dM = \frac{\widetilde{q}(\widetilde{M};t)}{\phi c} \tag{5}$$

is the strength of the source. Therefore the diffusion equation with a concentrated, point source of strength $\widetilde{q}(\widetilde{M};t)/(\phi c)$ at $\widetilde{M}$ is

$$Lp(M, \widetilde{M}; t) = \frac{\widetilde{q}(\widetilde{M};t)}{\phi c} \delta(M, \widetilde{M}). \tag{6}$$

If we also require that the action of the source be instantaneous at time $t = \widetilde{t}$, then for $\widetilde{t} \in T$, we must have

$$\int_T \int_\Omega f(M, \widetilde{M}; t, \widetilde{t}\,) dM dt = \frac{\widetilde{q}(\widetilde{M}; \widetilde{t}\,)}{\phi c}. \tag{7}$$

Thus, the corresponding source density is

$$f(M,\widetilde{M};t,\widetilde{t}\,)=\frac{\widetilde{q}(\widetilde{M};\widetilde{t}\,)}{\phi c}\delta(M,\widetilde{M})\delta(t,\widetilde{t}\,), \tag{8}$$

and the diffusion equation for an instantaneous and concentrated, point source of strength $\widetilde{q}(\widetilde{M};\widetilde{t}\,)/(\phi c)$ located at $\widetilde{M}$ and acting at $t=\widetilde{t}$ is

$$Lp(M,\widetilde{M};t,\widetilde{t}\,)=\frac{\widetilde{q}(\widetilde{M};\widetilde{t}\,)}{\phi c}\delta(M,\widetilde{M})\delta(t,\widetilde{t}). \tag{9}$$

We shall now extend the above ideas to define the *source density* for sources that are concentrated over a volume, on a surface, or along a line in Ω. This issue is of some importance to our discussion.

Let $\widetilde{\Omega}$ be a subdomain in Ω with $\widetilde{M}\in\widetilde{\Omega}$ and $\widetilde{T}$ be a time interval $\widetilde{T}$; $\{\widetilde{t}\,|\,t_1<\widetilde{t}<t_2\}$. Let us define a functional $I_{\widetilde{\Omega}\widetilde{T}}(M,\widetilde{M};t,\widetilde{t}\,)$ by

$$\begin{aligned}\langle I_{\widetilde{\Omega}\widetilde{T}},\phi\rangle &= \int_T\int_\Omega I_{\widetilde{\Omega}\widetilde{T}}(M,\widetilde{M};t,\widetilde{t}\,)\phi(M;t)dM\,dt\\ &=\frac{1}{a}\int_{\widetilde{T}}\int_{\widetilde{\Omega}}\phi(\widetilde{M};\widetilde{t}\,)d\widetilde{M}d\widetilde{t},\end{aligned} \tag{10}$$

that is, $I_{\widetilde{\Omega}\widetilde{T}}(M,\widetilde{M};t,\widetilde{t}\,)=1/a$ if $M\in\widetilde{\Omega}$ and $t\in\widetilde{T}$; otherwise $I_{\widetilde{\Omega}\widetilde{T}}(M,\widetilde{M};t,\widetilde{t}\,)=0$. It is clear that

$$I_{\widetilde{\Omega}\widetilde{T}}(M,\widetilde{M};t,\widetilde{t}\,)=\frac{1}{a}\int_{\widetilde{T}}\int_{\widetilde{\Omega}}\delta(M,\widetilde{M})\delta(t,\widetilde{t}\,)d\widetilde{M}d\widetilde{t}. \tag{11}$$

If $a=\int_{\widetilde{T}}\int_{\widetilde{\Omega}}d\widetilde{M}d\widetilde{t}$, then the functional $I_{\widetilde{\Omega}\widetilde{T}}$, when acting on a test function, $\phi(M;t)$, averages that test function over the volume of the domain $\widetilde{\Omega}$ and over the time interval $\widetilde{T}$. From (11), we visualize $I_{\widetilde{\Omega}\widetilde{T}}$ as the symbolic source density of a volumetric source of strength equal to unity that consists of instantaneous, point sources of strength $1/a$ distributed over the volume $\widetilde{\Omega}$ and the interval $\widetilde{T}$. If the strength of the source is $Q/(\phi c)$, then we define the source density of the volumetric source by

$$Q_{\widetilde{\Omega}\widetilde{T}}(M,\widetilde{M};t,\widetilde{t}\,)=\int_{\widetilde{T}}\int_{\widetilde{\Omega}}\frac{\widetilde{q}(\widetilde{M};\widetilde{t}\,)}{\phi c}\delta(M,\widetilde{M})\delta(t,\widetilde{t}\,)d\widetilde{M}d\widetilde{t}. \tag{12}$$

In (12), the strength of the point sources in $D=\Omega\times T$ is $\widetilde{q}(\widetilde{M};\widetilde{t}\,)/(\phi c)$. Noting that

$$\int_T\int_\Omega Q_{\widetilde{\Omega}\widetilde{T}}dM\,dt=\int_{\widetilde{T}}\int_{\widetilde{\Omega}}\frac{\widetilde{q}(\widetilde{M};\widetilde{t}\,)}{\phi c}d\widetilde{M}d\widetilde{t}=\frac{Q}{\phi c}, \tag{13}$$

it is also possible to interpret $\widetilde{q}(\widetilde{M};\widetilde{t}\,)/(\phi c)$ as the source density (subsequently referred to as density) of the point sources in $\widetilde{D}=\widetilde{\Omega}\times\widetilde{T}$. In this interpretation, if the volume of the fluid withdrawn from the porous medium through the source (the volume of $\widetilde{\Omega}$)

for a period of time $\widetilde{T}$ is Q, then the volume of fluid withdrawn per unit volume of the source per unit of time is $\widetilde{q}(\widetilde{M};\widetilde{t}\,)$. This interpretation is essential for relating the given production rate of an actual well with the density of the source representing that well. Densities of continuous and instantaneous sources can be obtained from (12). For example, if $\widetilde{T} = T$, then (12) yields the density of a continuous, volumetric source:

$$Q_{\widetilde{\Omega}c}(M,\widetilde{M};t) = \int_{\widetilde{\Omega}} \frac{\widetilde{q}(\widetilde{M};t)}{\phi c}\delta(M,\widetilde{M})d\widetilde{M}. \tag{14}$$

If $\widetilde{T}$ is an infinitesimal time-interval around $\widetilde{t}$, then (12) provides the density of an instantaneous, volumetric source

$$Q_{\widetilde{\Omega}i}(M,\widetilde{M};t,\widetilde{t}\,) = \int_{\widetilde{\Omega}} \frac{\widetilde{q}(\widetilde{M};\widetilde{t}\,)}{\phi c}\delta(M,\widetilde{M})\delta(t,\widetilde{t}\,)d\widetilde{M}. \tag{15}$$

Here, $\widetilde{q}(\widetilde{M};\widetilde{t}\,)/(\phi c)$ is the density of point sources in $\widetilde{\Omega}$; that is, $\widetilde{q}(\widetilde{M};\widetilde{t}\,)$ is the volume of the fluid withdrawn from unit volume of the source at time $t = \widetilde{t}$. *

Similarly, if the sources are distributed over the entire space-domain ($\widetilde{\Omega} = \Omega$) and are continuous ($\widetilde{T} = T$), then we obtain $Q_{\Omega c}(M;t) = \widetilde{q}(M;t)/\phi c$ as in §1.1. If $\widetilde{\Omega}$ is a spherical region of vanishingly small radius around $\widetilde{M}$, then $Q_{pc} = \widetilde{q}(\widetilde{M};t)\delta(M,\widetilde{M})/(\phi c)$ is the density of the continuous, point source and $Q_{pi} = \widetilde{q}(\widetilde{M};\widetilde{t}\,)\delta(M,\widetilde{M})\delta(t,\widetilde{t}\,)/(\phi c)$ is the density of the instantaneous, point source.**

The densities of surface and line sources can also be written in a form similar to that given by (12). If $\widetilde{q}(\widetilde{M};\widetilde{t}\,)/(\phi c)$ is the density of an instantaneous, point source in $\widetilde{D} = \widetilde{S} \times \widetilde{T}$ where $\widetilde{S} = \widetilde{\Omega}$ for a volumetric source, $\widetilde{S} = \widetilde{\Gamma}$ for a surface source, and $\widetilde{S} = \widetilde{C}$ for a line source, then

$$Q_{pi} = \frac{\widetilde{q}(\widetilde{M};\widetilde{t}\,)}{\phi c}\delta(M,\widetilde{M})\delta(t,\widetilde{t}\,) \tag{16}$$

is the density for an instantaneous, point source in $D = \Omega \times T$. Thus, letting $f(M,\widetilde{M};t,\widetilde{t}\,)$ denote the source density in D, we have

$$f(M,\widetilde{M};t,\widetilde{t}\,) = \int_{\widetilde{T}}\int_{\widetilde{S}} Q_{pi}(M,\widetilde{M};t,\widetilde{t}\,)d\widetilde{S}d\widetilde{t}, \tag{17}$$

---

* *(15) represents the appropriate limit of (12) if the source-strength remains constant while $\widetilde{T}$ becomes a vanishingly small time-interval around $\widetilde{t}$ (that is, the strength of the instantaneous source is the same as that of an impulsive source acting over a time period, $\widetilde{T}$). This procedure ensures that the superposition of instantaneous source densities given in (15) over a time interval $\widetilde{T}$ yields the impulsive source density given in (12).*

** *Here, we take the limit of (12) by keeping the source-strength constant and making $\widetilde{\Omega}$ vanish around $\widetilde{M}$. Similar to the discussion following (15), superposition of the point-source densities over the volume of $\widetilde{\Omega}$ yields the density for a volumetric source.*

and may write the diffusion equation as

$$Lp\left(M,\widetilde{M};t,\tilde{t}\,\right)=f\left(M,\widetilde{M};t,\tilde{t}\,\right);\quad \left(M,\widetilde{M};t,\tilde{t}\,\right)\in D. \tag{18}$$

Because we now work with impulsive and/or concentrated sources, we expect p to have singularities in Ω and T. We can, however, apply the Laplace transformation to (18) as in the classical sense (see Churchill [11] and LePage [12]) and obtain

$$\overline{L}\overline{p}(M,\widetilde{M};s,\tilde{t}\,)=\overline{f}(M,\widetilde{M};s,\tilde{t}\,)-p_i. \tag{19}$$

Here, $\overline{L}$ is defined in 1.1(18) and

$$\begin{aligned}\overline{f}(M,\widetilde{M};s,\tilde{t}\,)&=\int_{\widetilde{T}}\int_{\widetilde{S}}\frac{\widetilde{q}(\widetilde{M};\tilde{t}\,)}{\phi c}\delta(M,\widetilde{M})\exp(-s\tilde{t}\,)d\widetilde{S}d\tilde{t}\\&=\int_{\widetilde{T}}\int_{\widetilde{S}}\overline{Q}_{pi}(M,\widetilde{M};s,\tilde{t})d\widetilde{S}d\tilde{t}.\end{aligned} \tag{20}$$

For a point source at $\widetilde{M}$, we take the limit of (20), by keeping the strength of the source constant as $\widetilde{S}$ vanishes. This yields

$$\overline{f}(M,\widetilde{M};s,\tilde{t}\,)=\int_{\widetilde{T}}\overline{Q}_{pi}(M,\widetilde{M};s,\tilde{t}\,)d\tilde{t}. \tag{21}$$

Similarly, for an instantaneous source

$$\overline{f}(M,\widetilde{M};s,\tilde{t}\,)=\int_{\widetilde{S}}\overline{Q}_{pi}(M,\widetilde{M};s,\tilde{t}\,)d\widetilde{S}. \tag{22}$$

Note that for instantaneous sources acting at $t=0$, the initial value of p, p_i, can be discontinuous at the source location. These discontinuities, however, disappear after a very short period of time and thus $\lim_{t\to 0^+}p=p_i$ is continuous at all points except at the location of the source. Of particular interest is the density for a continuous source, the expression of which is given by

$$\begin{aligned}\overline{f}(M,\widetilde{M};s)&=\int_{T}\int_{\widetilde{S}}\frac{\widetilde{q}(\widetilde{M};t)}{\phi c}\delta(M,\widetilde{M})\exp(-st)d\widetilde{S}dt\\&=\int_{\widetilde{S}}\frac{\overline{\widetilde{q}}(\widetilde{M};s)}{\phi c}\delta(M,\widetilde{M})d\widetilde{S}.\end{aligned} \tag{23}$$

1.3. Integral representation of solutions to the diffusion equation; the fundamental solution

Our interest is in obtaining the integral representation of solutions to the diffusion equation (see, 1.2(19)),

$$\overline{L}\overline{p}(M_D, \widetilde{M}_D; s, \widetilde{t}_D) = \overline{f}(M_D, \widetilde{M}_D; s, \widetilde{t}_D) - p_i; \quad (M_D, \widetilde{M}_D; \widetilde{t}_D) \in D_D, \tag{1}$$

subject to one of the following boundary conditions,

$$\overline{p}|_{\Gamma_D} = \overline{g}, \tag{2}$$

$$\frac{\partial \overline{p}}{\partial n}|_{\Gamma_D} = \overline{h}, \tag{3}$$

or

$$\left(\frac{\partial \overline{p}}{\partial n} + \overline{p}\right)\Big|_{\Gamma_D} = \overline{r}, \tag{4}$$

where $\partial/\partial n$ denotes differentiation along the outward normal of the boundary, Γ_D. This objective can be accomplished by using the fundamental solution, $\overline{\gamma}(M_D, M'_D; s, 0)$, that corresponds to the operator $\overline{L}$ which satisfies

$$\overline{L}\overline{\gamma}(M_D, M'_D; s, 0) = -\delta(M_D, M'_D); \quad (M_D . M'_D; 0) \in D_D, \tag{5}$$

and the respective, homogeneous boundary-condition of the original problem ((2), (3), or (4) with $\overline{g} = \overline{h} = \overline{r} = 0$). As we discussed in §1.2, $\overline{\gamma}$ represents the Laplace transformation of the response of an instantaneous, point source of unit strength located at M'_D and acting at $t'_D = 0$ with $\gamma = 0$ for $t_D < 0$ (the casuality condition). We note that $\overline{L}$ is a self-adjoint elliptic operator and the existence of the fundamental solution is assured by the general theory of elliptic PDE's. We also note that *

$$\overline{\gamma}(M_{D1}, M_{D2}; s, 0) = \overline{\gamma}(M_{D2}, M_{D1}; s, 0); \tag{6}$$

---

* *To prove (6), we let* $u = \overline{\gamma}(M_D, M_{D1}; s, 0)$ *and* $v = \overline{\gamma}(M_D, M_{D2}; s, 0)$ *be the solutions for* $\overline{L}u = -\delta(M_D, M_{D1})$ *and* $\overline{L}v = -\delta(M_D, M_{D2})$*, respectively. Applying Green's identity to* $\overline{L}u$ *and* $\overline{L}v$ *and noting that* u *and* v *satisfy the homogeneous boundary-conditions, we may write*

$$\int_{\Omega_D} [\overline{\gamma}(M_D, M_{D2}; s, 0)\delta(M_D, M_{D1}) - \overline{\gamma}(M_D, M_{D1}; s, 0)\delta(M_D, M_{D2})]\, dM_D = 0,$$

which, by the fundamental property of the Dirac-delta function, gives (6).

that is, the fundamental solution is symmetric in space variables and satisfies *

$$\int_{\Omega_D} \overline{\gamma}(M_D, M'_D; s, 0) dM_D = \frac{1}{s}\left(1 + \int_{\Gamma_D} \frac{\partial \overline{\gamma}}{\partial n} d\Gamma_D\right). \tag{7}$$

The integral representation of the solution to (1) can be obtained by applying Green's identity to (1) and (5) as follows:

$$\int_{\Omega_D} \left(\overline{\gamma}\overline{L}\overline{p} - \overline{p}\overline{L}\overline{\gamma}\right) dM_D = \int_{\Gamma_D} \left(\overline{\gamma}\frac{\partial \overline{p}}{\partial n} - \overline{p}\frac{\partial \overline{\gamma}}{\partial n}\right) d\Gamma_D. \tag{8}$$

If we substitute the right-hand sides of (1) and (5) for $\overline{L}\overline{p}$ and $\overline{L}\overline{\gamma}$, respectively, in (8), interchange M_D and M'_D, and use the symmetry property of the fundamental solution (see (6)), (8) yields

$$\begin{aligned} \overline{p}(M_D, \widetilde{M}_D; s, \tilde{t}_D) = & \int_{\Omega_D} p_i \overline{\gamma}(M_D, M'_D; s, 0) dM'_D \\ & - \int_{\Omega_D} \overline{f}(M'_D, \widetilde{M}_D; s, \tilde{t}_D) \overline{\gamma}(M_D, M'_D; s, 0) dM'_D \\ & + \int_{\Gamma_D} \left(\overline{\gamma}\frac{\partial \overline{p}}{\partial n} - \overline{p}\frac{\partial \overline{\gamma}}{\partial n}\right) d\Gamma'_D. \end{aligned} \tag{9}$$

(9) is the general form of the integral representation of the solution to (1). In the following chapters, however, we will assume that the initial pressure distribution is uniform and $\overline{p}$ satisfies (2), (3), or (4) with $\overline{g} = \overline{r} = p_i/s$, $\overline{h} = 0$ as these assumptions hold true for most of the problems of interest in porous media. If we, then, use (7) and the condition that the fundamental solution satisfies the respective homogeneous boundary-condition of the original problem, we can recast (9) as

$$\overline{\Delta p}(M_D, \widetilde{M}_D; s, \tilde{t}_D) = \int_{\Omega_D} \overline{f}(M'_D, \widetilde{M}_D; s, \tilde{t}_D) \overline{\gamma}(M_D, M'_D; s, 0) dM'_D, \tag{10}$$

where

$$\overline{\Delta p} = \frac{p_i}{s} - \overline{p}. \tag{11}$$

(The expression in (10) could also be derived from (8) by replacing $\overline{p}$ by $\overline{\Delta p}$ and noting that the boundary integral in the right-hand side of (8) is zero, because $\overline{\gamma}$ and $\overline{\Delta p}$ satisfy the same homogeneous boundary-conditions.)

---

* *With the aid of Green's identity, this result is obtained from (5) as follows:*

$$\begin{aligned} \int_{\Omega_D} \overline{L}\overline{\gamma} dM_D &= \int_{\Omega_D} (\nabla_D^2 \overline{\gamma} - s\overline{\gamma}) dM_D \\ &= \int_{\Gamma_D} \frac{\partial \overline{\gamma}}{\partial n} d\Gamma_D - s \int_{\Omega_D} \overline{\gamma} dM_D = -1. \end{aligned}$$

Let us now scrutinize the result given in (10). Substituting the appropriate expression given in 1.2(20) for $\overline{f}$ in (10), we obtain

$$\begin{aligned}\overline{\Delta p}(M_D, \widetilde{M}_D; s, \widetilde{t}_D) &= \int_{\Omega_D} \int_{\widetilde{T}_D} \int_{\widetilde{S}_D} \frac{\widetilde{q}_D(\widetilde{M}_D; \widetilde{t}_D)}{\phi c} \delta(M'_D, \widetilde{M}_D) \exp(-s\widetilde{t}_D) d\widetilde{S}_D d\widetilde{t}_D \\ &\quad \overline{\gamma}(M_D, M'_D; s, 0) dM'_D \\ &= \int_{\widetilde{T}_D} \int_{\widetilde{S}_D} \frac{\widetilde{q}_D(\widetilde{M}_D; \widetilde{t}_D)}{\phi c} \exp(-s\widetilde{t}_D) \overline{\gamma}(M_D, \widetilde{M}_D; s, 0) d\widetilde{S}_D d\widetilde{t}_D.\end{aligned} \tag{12}$$

Here, $\overline{\Delta p}$ is the response of an impulsive and concentrated source and can be viewed as the suitable limit of the response to a continuous and distributed source. If we wish to express (12) in terms of the source density $\widetilde{q}/(\phi c)$ in $\widetilde{D} = \widetilde{S} \times \widetilde{T}$, then we use

$$\widetilde{q}_D = \frac{\ell^2}{\eta \ell^n} \widetilde{q}, \tag{13}$$

or for instantaneous sources

$$\widetilde{q}_D = \frac{1}{\ell^n} \widetilde{q}, \tag{14}$$

where $n = 0, 1, 2,$ or 3 for volume, plane, line or point sources, respectively. For example, for a continuous source (12) yields (see §1.2)

$$\begin{aligned}\overline{\Delta p}(M_D, \widetilde{M}_D; s) &= \int_{\widetilde{S}_D} \frac{\overline{\widetilde{q}}_D(\widetilde{M}_D; s)}{\phi c} \overline{\gamma}(M_D, \widetilde{M}_D; s) d\widetilde{S}_D \\ &= \frac{\mu}{k\ell} \int_{\widetilde{S}} \overline{\widetilde{q}}(\widetilde{M}_D; s) \overline{\gamma}(M_D, \widetilde{M}_D; s) d\widetilde{S}.\end{aligned} \tag{15}$$

If $\overline{q}$ is the volumetric rate at which fluid is extracted from the porous medium (Ω) through the source ($\widetilde{S}$), then $\int_{\widetilde{S}} \overline{\widetilde{q}} d\widetilde{S} = \overline{q}$. If $\widetilde{q}_D(\widetilde{M}_D; t_D)$ is known or can be computed by independent means, then (12) reduces the problems of unsteady flow in porous media to finding the fundamental solution (instantaneous, point-source solution of unit strength) that satisfies the same boundary conditions of the original problem. Therefore, in Chapter II we discuss how the fundamental solutions can be obtained subject to Dirichlet, Neumann, or mixed boundary conditions in the Cartesian and cylindrical coordinate systems. Chapter III is devoted to the discussion of the treatment of fluid extraction points, where we regard the wellbore as a source and evaluate (12) with $\widetilde{S}_D$ representing the wellbore geometry. Computational considerations and applications to fissured and layered porous media are discussed in Chapters IV and V, respectively.

References

1. Darcy, Henry: *Les Fontaines publiques de ville de Dijon*, Victor Delmont, Paris (1856).

2. Childs, E. C.: *The Physical Basis of Soil Water Phenomena*, Wiley–Interscience (1969).

3. Scorer, J. D. T. and Miller, F. G.: A Review of Rock Compressibility and its Relationship to Oil and Gas Recovery, Report IP 74-003 (1974), Institute of Petroleum, London.

4. Raghavan, R., Scorer, J. D. T., and Miller, F. G.: "A Study by Numerical Methods of the Effect of Pressure-Dependent Rock and Fluid Properties on Well Flow Tests," *Society of Petroleum Engineers Journal* (June 1972) 267–275.

5. Kirchhoff, G: "Vorlesungen über de theorie der Wärme," Barth, Leipzig (1894).

6. Cole, J. D.: "On a Quasi-Linear Parabolic Equation Occurring in Aerodynamics," *Quarterly of Applied Mathematics* (October 1951), 225–236.

7. Hopf, E.: "The Partial Differential Equation $u_t + uu_x = u_{xx}$," *Communication of Pure and Applied Mathematics* **3** (1950), 201–230.

8. Schwartz, L.: *Théorie des distributions* Vols. I and II. Actualitéx Scientifiques et Industrielles, Herman & Cie, Paris (1957, 1959).

9. Zemanian, A. H.: *Distribution Theory and Transform Analysis*, Dover, New York (1987).

10. Stakgold, I.: *Green's Functions and Boundary Value Problems*, Wiley–Interscience (1979).

11. Churchill, R. V.: *Operational Mathematics*, 3rd edition, McGraw-Hill Book Company (1972).

12. LePage, W. R.: *Complex Variables and the Laplace Transforms for Engineers*, Dover, New York (1961).

II. The Initial-Boundary-Value Problem

As noted in §1.3, the solutions for unsteady flow in porous media reduce to obtaining the fundamental solution. The fundamental solution will satisfy the homogeneous boundary conditions of the original problem. In this chapter, we establish the framework for examining three-dimensional flow in porous media by deriving fundamental solutions subject to Dirichlet, Neumann, or mixed boundary-conditions. Solutions are expressed in terms of the Laplace transformation and are derived by the well-known method of images.

2.1. Flow in an infinite medium

The fundamental solution, $\overline{\gamma}$, satisfies the Laplace transformation of the diffusion equation,

$$\overline{L}\overline{\gamma}(M_D, M_D'; s, 0) \equiv \nabla_D^2 \overline{\gamma} - s\overline{\gamma} = -\delta(M_D, M_D'); \quad (M_D, M_D'; 0) \in D_D. \tag{1}$$

As noted in §1.2 and §1.3, $\overline{\gamma}$ corresponds to the solution for an instantaneous, point source of unit strength that is located at M_D' and acts at $t_D' = 0$. For a point source located at the origin in an isotropic system, we may write (1) in spherical coordinates as follows:

$$\frac{1}{\rho_D^2}\frac{d}{d\rho_D}\left(\rho_D^2 \frac{d\overline{\gamma}}{d\rho_D}\right) - s\overline{\gamma} = 0 \quad \text{for } M_D \neq M_D', \tag{2}$$

where ρ_D is the radial coordinate. The solution of (2) yields the well-known Lord Kelvin's, point-source solution

$$\overline{\gamma} = \frac{\exp(-\rho_D\sqrt{s})}{4\pi\rho_D}. \tag{3}$$

In the following sections, we use (3) to derive the fundamental solutions for problems in Cartesian and cylindrical reference frames by the well-known method of images. (3) will be used after translation (to account for the fact that the source may not be at the origin), and the influence of anisotropy will be incorporated along the lines noted in §1.1.

2.2. Flow in linear (slab) porous media

We consider flow in a linear reservoir with boundaries at $z = 0$ and $z = z_e$. As noted previously, the fundamental solutions we derive here satisfy the homogeneous boundary conditions of the original problem. We consider a system with permeabilities, k_x, k_y and k_z in the x, y and z directions, respectively. We will assume that anisotropy is incorporated through the coordinate transformation discussed in Chapter I. The source is located at (x', y', z').

I. *Both ends of the porous solid are impermeable*

The fundamental solution, obtained by the method of images, is given by

$$\bar{\gamma} = \frac{1}{4\pi} \sum_{n=-\infty}^{+\infty} \left\{ \frac{\exp\left[-\sqrt{u}\,\sqrt{R_D^2 + (z_D - z'_D - 2nz_{eD})^2}\right]}{\sqrt{R_D^2 + (z_D - z'_D - 2nz_{eD})^2}} + \frac{\exp\left[-\sqrt{u}\,\sqrt{R_D^2 + (z_D + z'_D - 2nz_{eD})^2}\right]}{\sqrt{R_D^2 + (z_D + z'_D - 2nz_{eD})^2}} \right\}. \tag{1}$$

Here, $u = s$; as we shall see in Chapter V, this nomenclature permits us to use the development given here to derive solutions for naturally-fractured porous media by a trite change in nomenclature. In (1)

$$R_D^2 = (x_D - x'_D)^2 + (y_D - y'_D)^2, \tag{2}$$

$$x_D = \frac{x}{\ell}\sqrt{\frac{k}{k_x}}, \; y_D = \frac{y}{\ell}\sqrt{\frac{k}{k_y}}, \; z_D = \frac{z}{\ell}\sqrt{\frac{k}{k_z}} \tag{3}$$

and

$$z_{eD} = \frac{z_e}{\ell}\sqrt{\frac{k}{k_z}}, \tag{4}$$

(see §1.1, Remark 3). (1) can be recast into a more suitable form for computational purposes by using Poisson's summation formula (see Carslaw and Jaeger [1], p. 275) given by

$$\sum_{n=-\infty}^{+\infty} \exp\left[-\frac{(\xi - 2n\xi_e)^2}{4t_D}\right] = \frac{\sqrt{\pi t_D}}{\xi_e}\left[1 + 2\sum_{n=1}^{\infty} \exp\left(-\frac{n^2\pi^2 t_D}{\xi_e^2}\right)\cos n\pi\frac{\xi}{\xi_e}\right]. \tag{5}$$

After multiplying both sides of (5) by $t_D^{-3/2}\exp[-a^2/(4t_D)]$, where a is real and positive, the Laplace transformation with respect to t_D yields (see Ozkan [2], p. 17)

$$\sum_{n=-\infty}^{+\infty} \frac{\exp\left[-\sqrt{u}\,\sqrt{a^2 + (\xi - 2n\xi_e)^2}\right]}{\sqrt{a^2 + (\xi - 2n\xi_e)^2}} = \frac{1}{\xi_e}\left[K_0(a\sqrt{u}) + 2\sum_{n=1}^{\infty} K_0\left(a\sqrt{u + \frac{n^2\pi^2}{\xi_e^2}}\right)\cos n\pi\frac{\xi}{\xi_e}\right]. \tag{6}$$

If we use the summation formula given by (6), then the fundamental solution for a porous medium bounded by two impermeable planes at $z = 0$ and $z = z_e$ is

$$\bar{\gamma} = \frac{1}{2\pi z_{eD}} \left[K_0(R_D\sqrt{u}) + 2\sum_{n=1}^{\infty} K_0\left(R_D\sqrt{u + \frac{n^2\pi^2}{z_{eD}^2}}\right) \cos n\pi\frac{z_D}{z_{eD}} \cos n\pi\frac{z'_D}{z_{eD}}\right]. \tag{7}$$

In passing, we note that Poisson's summation formula given in (5) is equivalent to the transformation formula for *theta-functions*. In fact, all expressions derived in this section can be expressed in terms of theta-functions.

II. *Both ends of the porous solid are at the initial pressure*

Here, we assume that the boundaries are at a pressure equal to the initial pressure of the system. The fundamental solution in this case is given by

$$\bar{\gamma} = \frac{1}{4\pi} \sum_{n=-\infty}^{+\infty} \left\{ \frac{\exp\left[-\sqrt{u}\sqrt{R_D^2 + (z_D - z'_D - 2nz_{eD})^2}\right]}{\sqrt{R_D^2 + (z_D - z'_D - 2nz_{eD})^2}} - \frac{\exp\left[-\sqrt{u}\sqrt{R_D^2 + (z_D + z'_D - 2nz_{eD})^2}\right]}{\sqrt{R_D^2 + (z_D + z'_D - 2nz_{eD})^2}} \right\}, \tag{8}$$

which, by (6), becomes

$$\bar{\gamma} = \frac{1}{\pi z_{eD}} \left[\sum_{n=1}^{\infty} K_0\left(R_D\sqrt{u + \frac{n^2\pi^2}{z_{eD}^2}}\right) \sin n\pi\frac{z_D}{z_{eD}} \sin n\pi\frac{z'_D}{z_{eD}}\right]. \tag{9}$$

III. *One end of the porous solid is impermeable, the other end is at the initial pressure*

Let the boundary at $z = 0$ be impermeable and the boundary at $z = z_e$ be at a pressure equal to the initial pressure. For this case, the fundamental solution is given by

$$\bar{\gamma} = \frac{1}{4\pi} \sum_{n=-\infty}^{+\infty} (-1)^n \left\{ \frac{\exp\left[-\sqrt{u}\sqrt{R_D^2 + (z_D - z'_D - 2nz_{eD})^2}\right]}{\sqrt{R_D^2 + (z_D - z'_D - 2nz_{eD})^2}} + \frac{\exp\left[-\sqrt{u}\sqrt{R_D^2 + (z_D + z'_D - 2nz_{eD})^2}\right]}{\sqrt{R_D^2 + (z_D + z'_D - 2nz_{eD})^2}} \right\}. \tag{10}$$

Noting that

$$\sum_{k=-\infty}^{+\infty} (-1)^k \exp\left[-\frac{a(x_D - 2kx_{eD})^2}{4\xi}\right] = \sum_{k=-\infty}^{+\infty} \left\{ 2\exp\left[-\frac{a(x_D - 2k2x_{eD})^2}{4\xi}\right] - \exp\left[-\frac{a(x_D - 2kx_{eD})^2}{4\xi}\right]\right\}, \tag{11}$$

and using (6), we may recast (10) in the following form:

$$\overline{\gamma} = \frac{1}{\pi z_{eD}} \left\{ \sum_{n=1}^{\infty} K_0 \left[R_D \sqrt{u + \frac{(2n-1)^2\pi^2}{4z_{eD}^2}} \right] \cos(2n-1)\frac{\pi}{2}\frac{z_D}{z_{eD}} \cos(2n-1)\frac{\pi}{2}\frac{z'_D}{z_{eD}} \right\}. \tag{12}$$

2.3. Flow in cylindrical porous media

We assume that the porous medium is isotropic in the $x - y$ plane and let k_r denote the permeability in the $x - y$ plane. The vertical permeability will be represented by k_z. The bounding planes at $z = 0$ and $z = z_e$ may both be impermeable, both be at a constant (initial) pressure, or the plane at $z = 0$ may be impermeable while the plane at $z = z_e$ is at a constant (initial) pressure.

We consider two problems: flow in a finite cylinder with the boundary at $r = r_e$ considered to be impermeable or at a pressure equal to the initial pressure, and flow in a composite region with a change in properties at $r = a$.

I. *Flow in a finite cylinder*

The fundamental solution can be obtained in a formal way (see Carslaw and Jaeger [3], Chapter XIV). Considering the development in §2.2, however, one can readily write down the solutions merely by inspection. We derive the solutions in cylindrical coordinates where (r_D, θ, z_D) and (r'_D, θ', z'_D) are the coordinates of the points M_D and M'_D, respectively.

We seek a solution of

$$\frac{1}{r_D}\frac{\partial}{\partial r_D}\left(r_D \frac{\partial \overline{\gamma}}{\partial r_D}\right) + \frac{1}{r_D^2}\frac{\partial^2 \overline{\gamma}}{\partial \theta^2} + \frac{\partial^2 \overline{\gamma}}{\partial z_D^2} - u\overline{\gamma} = -\delta\left(M_D, M'_D; u, 0\right), \tag{1}$$

in the form

$$\overline{\gamma} = P + G. \tag{2}$$

In (2), P is a solution of (1) that satisfies the conditions at the location of the source, M'_D, and also at the boundaries $z = 0$ and $z = z_e$. It is obvious that one of the solutions given in §2.2 for the appropriate boundary conditions at $z = 0$ and $z = z_e$ can be used for P. In addition to satisfying the boundary conditions at $z = 0$ and $z = z_e$, G is chosen such that $P + G$ satisfies the boundary condition at $r = r_e$ and the contribution of G to the flux vanishes as $M_D \to M'_D$. This procedure was used by Muskat [4] in the study of steady flow in porous media.

Let us first consider the case wherein the boundary conditions at $z_D = 0$, $z_D = z_{eD}$ and $r_D = r_{eD}$ are given, respectively, by

$$\frac{\partial \overline{\gamma}}{\partial z_D}\Big|_{z_D = 0, z_{eD}} = 0, \tag{3}$$

and

$$\frac{\partial \overline{\gamma}}{\partial r_D}\Big|_{r_D = r_{eD}} = 0. \tag{4}$$

In accordance with the boundary conditions given by (3), we choose P as the solution given by 2.2(7). If we use the addition theorem for the Bessel function $K_0(aR_D)$ (see Carslaw and Jaeger [5], p. 377) given by

$$\left.\begin{aligned} K_0(aR_D) &= \sum_{k=-\infty}^{+\infty} I_k(ar_D)K_k(ar'_D)\cos k(\theta-\theta');\ r_D < r'_D \\ K_0(aR_D) &= \sum_{k=-\infty}^{+\infty} I_k(ar'_D)K_k(ar_D)\cos k(\theta-\theta');\ r_D > r'_D \end{aligned}\right\}, \tag{5}$$

where $R_D^2 = r_D^2 + r_D'^2 - 2r_D r'_D \cos(\theta-\theta')$, we can write P as

$$\begin{aligned} P = \frac{1}{2\pi z_{eD}} \Bigg[& \sum_{k=-\infty}^{+\infty} I_k\left(\sqrt{u}r_D\right) K_k\left(\sqrt{u}r'_D\right) \cos k(\theta-\theta') \\ & + 2\sum_{n=1}^{\infty} \cos n\pi \frac{z_D}{z_{eD}} \cos n\pi \frac{z'_D}{z_{eD}} \\ & \sum_{k=-\infty}^{+\infty} I_k\left(\sqrt{u+\frac{n^2\pi^2}{z_{eD}^2}}r_D\right) K_k\left(\sqrt{u+\frac{n^2\pi^2}{z_{eD}^2}}r'_D\right) \cos k(\theta-\theta') \Bigg] \end{aligned} \tag{6}$$

for $r_D < r'_D$. For $r_D > r'_D$, interchange r_D and r'_D in (6). Let

$$\begin{aligned} G = \frac{1}{2\pi z_{eD}} \Bigg[& \sum_{k=-\infty}^{+\infty} a_k I_k\left(\sqrt{u}r_D\right) \cos k(\theta-\theta') \\ & + 2\sum_{n=1}^{\infty} \cos n\pi \frac{z_D}{z_{eD}} \cos n\pi \frac{z'_D}{z_{eD}} \sum_{k=-\infty}^{+\infty} b_k I_k\left(\sqrt{u+\frac{n^2\pi^2}{z_{eD}^2}}r_D\right) \cos k(\theta-\theta') \Bigg]. \end{aligned} \tag{7}$$

This choice satisfies the condition given in (3) and the requirement that the contribution of G to the flux $\to 0$ as $M_D \to M'_D$. If a_k and b_k in (7) are now chosen as

$$a_k = -\frac{I_k\left(\sqrt{u}r'_D\right) K'_k\left(\sqrt{u}r_{eD}\right)}{I'_k\left(\sqrt{u}r_{eD}\right)} \tag{8}$$

and

$$b_k = -\frac{I_k\left(\sqrt{u+\frac{n^2\pi^2}{z_{eD}^2}}r'_D\right) K'_k\left(\sqrt{u+\frac{n^2\pi^2}{z_{eD}^2}}r_{eD}\right)}{I'_k\left(\sqrt{u+\frac{n^2\pi^2}{z_{eD}^2}}r_{eD}\right)}, \tag{9}$$

respectively, then $P+G$ satisfies (4), the condition at r_{eD}, and therefore the fundamental solution is given by

$$\overline{\gamma} = \frac{1}{2\pi z_{eD}} \Bigg\{ K_0(R_D\sqrt{u}) - \sum_{k=-\infty}^{+\infty} I_k\left(\sqrt{u}r_D\right) \frac{I_k(\sqrt{u}r'_D)\, K'_k(\sqrt{u}r_{eD})}{I'_k(\sqrt{u}r_{eD})} \cos k(\theta-\theta') + 2\sum_{n=1}^{\infty} \cos n\pi\frac{z_D}{z_{eD}} \cos n\pi\frac{z'_D}{z_{eD}} \left[K_0\left(R_D\sqrt{u+\frac{n^2\pi^2}{z_{eD}^2}}\right) - \sum_{k=-\infty}^{+\infty} I_k\left(\sqrt{u+\frac{n^2\pi^2}{z_{eD}^2}}r_D\right) \frac{I_k\left(\sqrt{u+\frac{n^2\pi^2}{z_{eD}^2}}r'_D\right) K'_k\left(\sqrt{u+\frac{n^2\pi^2}{z_{eD}^2}}r_{eD}\right)}{I'_k\left(\sqrt{u+\frac{n^2\pi^2}{z_{eD}^2}}r_{eD}\right)} \cos k(\theta-\theta') \right] \Bigg\}. \tag{10}$$

If, in the above system, we replace the boundary condition given by (4) with

$$\overline{\gamma}\,\Big|_{r_D=r_{eD}} = 0, \tag{11}$$

then the function P would still be given by the right-hand side of 2.2(7), but the function G would have to be chosen as

$$G = -\frac{1}{2\pi z_{eD}} \left[\sum_{k=-\infty}^{+\infty} I_k\left(\sqrt{u}r_D\right) \frac{I_k(\sqrt{u}r'_D)\, K_k(\sqrt{u}r_{eD})}{I_k(\sqrt{u}r_{eD})} \cos k(\theta-\theta') + 2\sum_{n=1}^{\infty} \cos n\pi\frac{z_D}{z_{eD}} \cos n\pi\frac{z'_D}{z_{eD}} \sum_{k=-\infty}^{+\infty} I_k\left(\sqrt{u+\frac{n^2\pi^2}{z_{eD}^2}}r_D\right) \frac{I_k\left(\sqrt{u+\frac{n^2\pi^2}{z_{eD}^2}}r'_D\right) K_k\left(\sqrt{u+\frac{n^2\pi^2}{z_{eD}^2}}r_{eD}\right)}{I_k\left(\sqrt{u+\frac{n^2\pi^2}{z_{eD}^2}}r_{eD}\right)} \cos k(\theta-\theta') \right]. \tag{12}$$

Therefore, the solution satisfying the boundary condition given by (11) would be

$$
\begin{aligned}
\overline{\gamma} = & \frac{1}{2\pi z_{eD}} \Bigg\{ K_0(R_D\sqrt{u}) \\
& - \sum_{k=-\infty}^{+\infty} I_k\left(\sqrt{u}r_D\right) \frac{I_k\left(\sqrt{u}r'_D\right) K_k\left(\sqrt{u}r_{eD}\right)}{I_k\left(\sqrt{u}r_{eD}\right)} \cos k(\theta-\theta') \\
& + 2\sum_{n=1}^{\infty} \cos n\pi \frac{z_D}{z_{eD}} \cos n\pi \frac{z'_D}{z_{eD}} \Bigg[K_0\left(R_D\sqrt{u+\frac{n^2\pi^2}{z_{eD}^2}}\right) \\
& - \sum_{k=-\infty}^{+\infty} I_k\left(\sqrt{u+\frac{n^2\pi^2}{z_{eD}^2}} r_D\right) \\
& \frac{I_k\left(\sqrt{u+\frac{n^2\pi^2}{z_{eD}^2}} r'_D\right) K_k\left(\sqrt{u+\frac{n^2\pi^2}{z_{eD}^2}} r_{eD}\right)}{I_k\left(\sqrt{u+\frac{n^2\pi^2}{z_{eD}^2}} r_{eD}\right)} \cos k(\theta-\theta') \Bigg] \Bigg\}.
\end{aligned} \tag{13}
$$

In general, after choosing the function P as one of the solutions given by 2.2(7), 2.2(9), or 2.2(12) for the appropriate boundary conditions at $z = 0$ and $z = z_e$, the function G can be obtained by replacing the $K_0(aR_D)$ terms in the function P with

$$
-\sum_{k=-\infty}^{+\infty} I_k(ar_D)\frac{I_k(ar'_D)K'_k(ar_{eD})}{I'_k(ar_{eD})} \cos k(\theta-\theta') \quad \text{if } \left.\frac{\partial\overline{\gamma}}{\partial r_D}\right|_{r_D=r_{eD}} = 0, \tag{14}
$$

and with

$$
-\sum_{k=-\infty}^{+\infty} I_k(ar_D)\frac{I_k(ar'_D)K_k(ar_{eD})}{I_k(ar_{eD})} \cos k(\theta-\theta') \quad \text{if } \left.\overline{\gamma}\right|_{r_D=r_{eD}} = 0. \tag{15}
$$

For $r'_D = 0$, the expressions given in (14) and (15) reduce, respectively, to

$$
I_0(ar_D)\frac{K_1(ar_{eD})}{I_1(ar_{eD})} \quad \text{if } \left.\frac{\partial\overline{\gamma}}{\partial r_D}\right|_{r_D=r_{eD}} = 0 \tag{16}
$$

and

$$
-I_0(ar_D)\frac{K_0(ar_{eD})}{I_0(ar_{eD})} \quad \text{if } \left.\overline{\gamma}\right|_{r_D=r_{eD}} = 0. \tag{17}
$$

II. *Composite regions*

For illustration, we assume that the planes $z = 0$ and $z = z_e$ are impermeable. The inner region extends over a region $0 \leq r < a$ (Region 1) and the outer region, Region 2, with properties distinct from Region 1, is assumed to be infinite in extent. The source is located at r', θ', z' and $r' < a$, that is, the source is in Region 1. Solutions given here may be readily extended to other circumstances (source in the outer region, Region 2 is bounded, etc.)

We decouple the problems for Region 1 and Region 2. The fundamental solution, γ_1, for Region 1 satisfies

$$\frac{1}{r_D}\frac{\partial}{\partial r_D}\left(r_D\frac{\partial\overline{\gamma}_1}{\partial r_D}\right)+\frac{1}{r_D^2}\frac{\partial^2\overline{\gamma}_1}{\partial\theta^2}+\frac{\partial^2\overline{\gamma}_1}{\partial z_D^2}-u\overline{\gamma}_1=-\delta(M_D,M'_D;u,0). \tag{18}$$

If we write $\overline{\gamma}_1 = P + G$ and follow the development in I, then the solution for Region 1 is given by

$$\begin{aligned}\overline{\gamma}_1 = \frac{1}{2\pi z_{eD}}\Bigg[&\sum_{k=-\infty}^{+\infty} I_k\left(\sqrt{u}r'_D\right)K_k\left(\sqrt{u}r_D\right)\cos k(\theta-\theta')\\
&+2\sum_{n=1}^{\infty}\cos n\pi\frac{z_D}{z_{eD}}\cos n\pi\frac{z'_D}{z_{eD}}\sum_{k=-\infty}^{+\infty} I_k\left(\sqrt{u+\frac{n^2\pi^2}{z_{eD}^2}}r'_D\right)\\
&K_k\left(\sqrt{u+\frac{n^2\pi^2}{z_{eD}^2}}r_D\right)\cos k(\theta-\theta')\\
&+\sum_{k=-\infty}^{+\infty} a_{k0}I_k\left(\sqrt{u}r_D\right)\cos k(\theta-\theta')\\
&+2\sum_{n=1}^{\infty}\cos n\pi\frac{z_D}{z_{eD}}\cos n\pi\frac{z'_D}{z_{eD}}\sum_{k=-\infty}^{+\infty} a_{kn}I_k\left(\sqrt{u+\frac{n^2\pi^2}{z_{eD}^2}}r_D\right)\cos k(\theta-\theta')\Bigg],\end{aligned} \tag{19}$$

for $r_D > r'_D$. For $r_D < r'_D$ we interchange r_D and r'_D.

We now consider an infinite porous medium having the properties of Region 2. The counterpart of the fundamental solution, γ_1, γ_2, for this region, satisfies

$$\frac{1}{r_D}\frac{\partial}{\partial r_D}\left(r_D\frac{\partial\overline{\gamma}_2}{\partial r_D}\right)+\frac{1}{r_D^2}\frac{\partial^2\overline{\gamma}_2}{\partial\theta^2}+\frac{\partial^2\overline{\gamma}_2}{\partial\tilde{z}_D^2}-\tilde{u}\overline{\gamma}_2=0, \tag{20}$$

where the normalized variables are based on the properties of Region 1,

$$\tilde{z}_D=\sqrt{\frac{k_{r2}k_{z1}}{k_{r1}k_{z2}}}z_D, \tag{21}$$

$$\tilde{u}=\eta_{rD}u, \tag{22}$$

and

$$\eta_{rD}=\frac{\eta_{r1}}{\eta_{r2}}=\frac{k_1\phi_2c_2}{k_2\phi_1c_1}. \tag{23}$$

The solution $\overline{\gamma}_2$ is given by

$$\begin{aligned}\overline{\gamma}_2=\frac{1}{2\pi\tilde{z}_{eD}}\Bigg[&\sum_{k=-\infty}^{+\infty} b_{k0}K_k\left(\sqrt{\tilde{u}}r_D\right)\cos k(\theta-\theta')\\
&+2\sum_{n=1}^{\infty}\cos n\pi\frac{\tilde{z}_D}{\tilde{z}_{eD}}\cos n\pi\frac{\tilde{z}'_D}{\tilde{z}_{eD}}\sum_{k=-\infty}^{+\infty} b_{kn}K_k\left(\sqrt{\tilde{u}+\frac{n^2\pi^2}{\tilde{z}_{eD}^2}}r_D\right)\cos k(\theta-\theta')\Bigg],\end{aligned} \tag{24}$$

where

$$\widetilde{z}_{eD} = \sqrt{\frac{k_{r2}k_{z1}}{k_{r1}k_{z2}}}\, z_{eD}. \tag{25}$$

The coefficients a_{kn} and b_{kn} in (19) and (24) can now be found by coupling γ_1 and γ_2 at $r_D = a_D$. The coupling conditions are given by

$$\overline{\gamma}_1 = \overline{\gamma}_2 \quad \text{at } r_D = a_D, \tag{26}$$

and

$$\frac{k_{r1}}{\mu}\frac{\partial \overline{\gamma}_1}{\partial r_D} = \frac{k_{r2}}{\mu}\frac{\partial \overline{\gamma}_2}{\partial r_D} \quad \text{at } r_D = a_D. \tag{27}$$

Because the terms in the summations in (19) and (24) are independent of one another, (26) and (27) must hold for each value of the subscripts k and n. Thus, we obtain

$$a_{kn} = -I_k\left(\sqrt{u + \frac{n^2\pi^2}{z_{eD}^2}}\, r'_D\right)\frac{\Omega_{kn}}{\Psi_{kn}}, \tag{28}$$

and

$$b_{kn} = \frac{\widetilde{z}_{eD}}{z_{eD}}\frac{\lambda_{rD}}{a_D}\frac{I_k\left(\sqrt{u + \frac{n^2\pi^2}{z_{eD}^2}}\, r'_D\right)}{\Psi_{kn}}, \tag{29}$$

where

$$\begin{aligned}\Omega_{kn} =& \lambda_{rD}\sqrt{u + \frac{n^2\pi^2}{z_{eD}^2}}\, K_k\left(\sqrt{\widetilde{u} + \frac{n^2\pi^2}{\widetilde{z}_{eD}^2}}\, a_D\right) K'_k\left(\sqrt{u + \frac{n^2\pi^2}{z_{eD}^2}}\, a_D\right) \\ & - \sqrt{\widetilde{u} + \frac{n^2\pi^2}{\widetilde{z}_{eD}^2}}\, K_k\left(\sqrt{u + \frac{n^2\pi^2}{z_{eD}^2}}\, a_D\right) K'_k\left(\sqrt{\widetilde{u} + \frac{n^2\pi^2}{\widetilde{z}_{eD}^2}}\, a_D\right),\end{aligned} \tag{30}$$

$$\begin{aligned}\Psi_{kn} =& \lambda_{rD}\sqrt{u + \frac{n^2\pi^2}{z_{eD}^2}}\, I'_k\left(\sqrt{u + \frac{n^2\pi^2}{z_{eD}^2}}\, a_D\right) K_k\left(\sqrt{\widetilde{u} + \frac{n^2\pi^2}{\widetilde{z}_{eD}^2}}\, a_D\right) \\ & - \sqrt{\widetilde{u} + \frac{n^2\pi^2}{\widetilde{z}_{eD}^2}}\, I_k\left(\sqrt{u + \frac{n^2\pi^2}{z_{eD}^2}}\, a_D\right) K'_k\left(\sqrt{\widetilde{u} + \frac{n^2\pi^2}{\widetilde{z}_{eD}^2}}\, a_D\right),\end{aligned} \tag{31}$$

and

$$\lambda_{rD} = \frac{k_{r1}}{k_{r2}}. \tag{32}$$

In obtaining (28) and (29), we have used the following Wronskian relation of Bessel functions (see Watson [6]):

$$W\{I_\nu(z), K_\nu(z)\} = I_\nu(z)K'_\nu(z) - I'_\nu(z)K_\nu(z) = -\frac{1}{z}. \tag{33}$$

Using (28) and (29), we now write the following expressions for $\overline{\gamma}_1$ and $\overline{\gamma}_2$ when $r_D > r'_D$:

$$\overline{\gamma}_1 = \frac{1}{2\pi z_{eD}} \left(\sum_{k=-\infty}^{+\infty} S_{k0} + 2 \sum_{n=1}^{\infty} \cos n\pi \frac{z_D}{z_{eD}} \cos n\pi \frac{z'_D}{z_{eD}} \sum_{k=-\infty}^{+\infty} S_{kn} \right), \tag{34}$$

and

$$\overline{\gamma}_2 = \frac{\lambda_{rD}}{2\pi z_{eD} a_D} \left(\sum_{k=-\infty}^{+\infty} R_{k0} + 2 \sum_{n=1}^{\infty} \cos n\pi \frac{\widetilde{z}_D}{\widetilde{z}_{eD}} \cos n\pi \frac{\widetilde{z}'_D}{\widetilde{z}_{eD}} \sum_{k=-\infty}^{+\infty} R_{kn} \right), \tag{35}$$

where

$$\begin{aligned} S_{kn} =& I_k \left(\sqrt{u + \frac{n^2\pi^2}{z_{eD}^2}} r'_D \right) \left[K_k \left(\sqrt{u + \frac{n^2\pi^2}{z_{eD}^2}} r_D \right) \right. \\ & \left. - \frac{\Omega_{kn}}{\Psi_{kn}} I_k \left(\sqrt{u + \frac{n^2\pi^2}{z_{eD}^2}} r_D \right) \right] \cos k(\theta - \theta'), \end{aligned} \tag{36}$$

$$R_{kn} = \frac{1}{\Psi_{kn}} I_k \left(\sqrt{u + \frac{n^2\pi^2}{z_{eD}^2}} r'_D \right) K_k \left(\sqrt{\widetilde{u} + \frac{n^2\pi^2}{\widetilde{z}_{eD}^2}} r_D \right) \cos k(\theta - \theta'). \tag{37}$$

When $r_D < r'_D$, we interchange r_D and r'_D in (34).

To obtain the pressure drop for a volumetric, plane, line, or point source, we use $\overline{\gamma}_1$ and $\overline{\gamma}_2$ given in (34) and (35) in 1.3(15). This yields

$$\overline{\Delta p}(r_D < a_D) = \frac{\mu}{k_1 \ell} \int_{\widetilde{S}} \overline{\widetilde{q}}\, \overline{\gamma}_1 d\widetilde{S} \tag{38}$$

and

$$\overline{\Delta p}(r_D > a_D) = \frac{\mu}{k_1 \ell} \int_{\widetilde{S}} \overline{\widetilde{q}}\, \overline{\gamma}_2 d\widetilde{S}. \tag{39}$$

2.4. Flow in rectangular parallelepipeds

We consider the fundamental solutions for flow in rectangular parallelepipeds. As in §2.2, combinations of boundary conditions are examined. Although obtaining the fundamental solutions in a rectangular parallelepiped by using the method of images is fairly easy, recasting the resulting expressions in a more manageable form for computational purposes is not straightforward. A triple Fourier series needs to be evaluated. The triple sums, however, can be readily reduced to double sums by the procedure outlined here. For two-dimensional problems, the triple infinite-series is reduced to the computation of a single infinite-series, and for three-dimensional problems, a double infinite-series needs to be computed.

The examples given below permit us to derive solutions for all combinations of boundary conditions. Let the porous medium occupy the region $0 < x < x_e$, $0 < y <$

y_e, and $0 < z < z_e$, and the source be located at x', y', z'. The assumptions noted in §2.2 also apply here.

I. *All boundaries of the porous solid are impermeable*

The fundamental solution is obtained by applying the method of images to the fundamental solution for an infinite medium (2.1(3)) and is given by

$$\begin{aligned}\overline{\gamma} = & \frac{1}{4\pi} \sum_{k=-\infty}^{+\infty} \sum_{m=-\infty}^{+\infty} \sum_{n=-\infty}^{+\infty} (S_{1,1,1} + S_{2,1,1} + S_{1,2,1} + S_{2,2,1} \\ & + S_{1,1,2} + S_{2,1,2} + S_{1,2,2} + S_{2,2,2}) .\end{aligned} \tag{1}$$

Here, we have defined

$$S_{i,j,l} = \frac{\exp\left[-\sqrt{u}\sqrt{(\tilde{x}_{Di} - 2kx_{eD})^2 + (\tilde{y}_{Dj} - 2my_{eD})^2 + (\tilde{z}_{Dl} - 2nz_{eD})^2}\right]}{\sqrt{(\tilde{x}_{Di} - 2kx_{eD})^2 + (\tilde{y}_{Dj} - 2my_{eD})^2 + (\tilde{z}_{Dl} - 2nz_{eD})^2}}, \tag{2}$$

for $i, j, l = 1$ or 2, where

$$\begin{aligned} \tilde{x}_{D1} &= x_D - x'_D \\ \tilde{x}_{D2} &= x_D + x'_D \\ \tilde{y}_{D1} &= y_D - y'_D \\ \tilde{y}_{D2} &= y_D + y'_D \\ \tilde{z}_{D1} &= z_D - z'_D \end{aligned} \tag{3}$$

and

$$\tilde{z}_{D2} = z_D + z'_D .$$

Physically, each $S_{i,j,l}$ represents the contribution of a given instantaneous, point source in an infinite array of sources.

We now proceed to rewrite the right-hand side of (1) in a form suitable for computational purposes. Let T_S represent one of the triple sums; that is,

$$T_S = \sum_{k=-\infty}^{+\infty} \sum_{m=-\infty}^{+\infty} \sum_{n=-\infty}^{+\infty} S, \tag{4}$$

where

$$S = \frac{\exp\left[-\sqrt{u}\,\sqrt{(\tilde{x}_D - 2kx_{eD})^2 + (\tilde{y}_D - 2my_{eD})^2 + (\tilde{z}_D - 2nz_{eD})^2}\right]}{\sqrt{(\tilde{x}_D - 2kx_{eD})^2 + (\tilde{y}_D - 2my_{eD})^2 + (\tilde{z}_D - 2nz_{eD})^2}}. \tag{5}$$

Using Poisson's summation formula given in 2.2(6) and the integral representation of the Macdonald function, $K_0(z)$, given by

$$K_0(z) = \frac{1}{2}\int_0^{\infty} \exp\left(-\xi - \frac{z^2}{4\xi}\right)\frac{d\xi}{\xi}; \quad \mathcal{R}e\ z^2) > 0, \tag{6}$$

we obtain

$$
\begin{aligned}
T_S = \frac{1}{2z_{eD}} \Bigg\{ \int_0^{\infty} \exp(-\xi) \sum_{k=-\infty}^{+\infty} \exp\left[-\frac{u(\widetilde{x}_D - 2kx_{eD})^2}{4\xi}\right] \\
\sum_{m=-\infty}^{+\infty} \exp\left[-\frac{u(\widetilde{y}_D - 2my_{eD})^2}{4\xi}\right] \frac{d\xi}{\xi} \\
+ 2\sum_{n=1}^{+\infty} \cos n\pi \frac{\widetilde{z}_D}{z_{eD}} \int_0^{\infty} \exp(-\xi) \sum_{k=-\infty}^{+\infty} \exp\left[-\frac{\left(u + \frac{n^2\pi^2}{z_{eD}^2}\right)(\widetilde{x}_D - 2kx_{eD})^2}{4\xi}\right] \\
\sum_{m=-\infty}^{+\infty} \exp\left[-\frac{\left(u + \frac{n^2\pi^2}{z_{eD}^2}\right)(\widetilde{y}_D - 2my_{eD})^2}{4\xi}\right] \frac{d\xi}{\xi} \Bigg\}.
\end{aligned}
\tag{7}
$$

Let T_1 be the second integral in (7). If we use the version of Poisson's summation formula given in 2.2(5) and then integrate, we obtain

$$
\begin{aligned}
T_1 = \frac{\pi}{ax_{eD}y_{eD}} \Bigg[1 + 2\sum_{k=1}^{\infty} \left(\frac{1}{1 + \frac{k^2\pi^2}{ax_{eD}^2}}\right) \cos k\pi \frac{\widetilde{x}_D}{x_{eD}} \\
+ 2\sum_{m=1}^{\infty} \left(\frac{1}{1 + \frac{m^2\pi^2}{ay_{eD}^2}}\right) \cos m\pi \frac{\widetilde{y}_D}{y_{eD}} \\
+4\sum_{k=1}^{\infty} \cos k\pi \frac{\widetilde{x}_D}{x_{eD}} \sum_{m=1}^{\infty} \left(\frac{1}{1 + \frac{k^2\pi^2}{ax_{eD}^2} + \frac{m^2\pi^2}{ay_{eD}^2}}\right) \cos m\pi \frac{\widetilde{y}_D}{y_{eD}} \Bigg],
\end{aligned}
\tag{8}
$$

where

$$
a = u + \frac{n^2\pi^2}{z_{eD}^2}. \tag{9}
$$

Because (see Gradshteyn and Ryzhik [7], 1.445-2)

$$
\sum_{k=1}^{\infty} \frac{\cos k\pi x}{k^2 + a^2} = \frac{\pi}{2a} \frac{ch[a\pi(1-x)]}{sh(a\pi)} - \frac{1}{2a^2}; \quad [0 \le x \le 2], \tag{10}
$$

where $ch(x)$ and $sh(x)$ represent the hyperbolic cosine of x and hyperbolic sine of x, respectively, T_1 is given by

$$
\begin{aligned}
T_1 = \frac{\pi}{\sqrt{a}x_{eD}} \frac{ch\sqrt{a}(y_{eD} - \widetilde{y}_D)}{sh\sqrt{a}y_{eD}} \\
+ \frac{2\pi}{x_{eD}} \sum_{k=1}^{\infty} \frac{\cos k\pi \frac{\widetilde{x}_D}{x_{eD}}}{\sqrt{a + \frac{\pi^2 k^2}{x_{eD}^2}}} \frac{ch\sqrt{a + \frac{\pi^2 k^2}{x_{eD}^2}}(y_{eD} - \widetilde{y}_D)}{sh\sqrt{a + \frac{\pi^2 k^2}{x_{eD}^2}}\, y_{eD}}.
\end{aligned}
\tag{11}
$$

Thus, (4) becomes

$$\sum_{k=-\infty}^{+\infty}\sum_{m=-\infty}^{+\infty}\sum_{n=-\infty}^{+\infty} S = \frac{\pi}{2x_{eD}z_{eD}}\left\{\frac{ch\sqrt{u}(y_{eD}-|\tilde{y}_D|)}{\sqrt{u}\ sh\sqrt{u}\ y_{eD}}\right.$$
$$+2\sum_{k=1}^{\infty}\cos k\pi\frac{\tilde{x}_D}{x_{eD}}\frac{ch\sqrt{u+\frac{\pi^2k^2}{x_{eD}^2}}\,(y_{eD}-|\tilde{y}_D|)}{\sqrt{u+\frac{\pi^2k^2}{x_{eD}^2}}\ sh\sqrt{u+\frac{\pi^2k^2}{x_{eD}^2}}\ y_{eD}}$$
$$+2\sum_{n=1}^{\infty}\cos n\pi\frac{\tilde{z}_D}{z_{eD}}\left[\frac{ch\sqrt{u+\frac{n^2\pi^2}{z_{eD}^2}}\,(y_{eD}-|\tilde{y}_D|)}{\sqrt{u+\frac{n^2\pi^2}{z_{eD}^2}}sh\sqrt{u+\frac{n^2\pi^2}{z_{eD}^2}}\ y_{eD}}\right.$$
$$\left.\left.+2\sum_{k=1}^{\infty}\cos k\pi\frac{\tilde{x}_D}{x_{eD}}\frac{ch\sqrt{u+\frac{k^2\pi^2}{x_{eD}^2}+\frac{n^2\pi^2}{z_{eD}^2}}\,(y_{eD}-|\tilde{y}_D|)}{\sqrt{u+\frac{k^2\pi^2}{x_{eD}^2}+\frac{n^2\pi^2}{z_{eD}^2}}\ sh\sqrt{u+\frac{k^2\pi^2}{x_{eD}^2}+\frac{n^2\pi^2}{z_{eD}^2}}\ y_{eD}}\right]\right\}. \tag{12}$$

We have reduced the triple infinite-series to expressions involving a double infinite-series. This procedure can be used to evaluate all eight triple-summation terms in (1). Thus, the fundamental solution for a porous solid in which all boundaries are impermeable is

$$\bar{\gamma} = \frac{1}{2x_{eD}z_{eD}}\left\{\frac{ch\sqrt{u}(y_{eD}-|\tilde{y}_{D1}|)+ch\sqrt{u}(y_{eD}-\tilde{y}_{D2})}{\sqrt{u}sh\sqrt{u}y_{eD}}\right.$$
$$+2\sum_{k=1}^{\infty}\cos k\pi\frac{x_D}{x_{eD}}\cos k\pi\frac{x'_D}{x_{eD}}$$
$$\frac{ch\sqrt{u+\frac{k^2\pi^2}{x_{eD}^2}}\,(y_{eD}-|\tilde{y}_{D1}|)+ch\sqrt{u+\frac{k^2\pi^2}{x_{eD}^2}}(y_{eD}-\tilde{y}_{D2})}{\sqrt{u+\frac{k^2\pi^2}{x_{eD}^2}}sh\sqrt{u+\frac{k^2\pi^2}{x_{eD}^2}}y_{eD}}$$
$$+2\sum_{n=1}^{\infty}\cos n\pi\frac{z_D}{z_{eD}}\cos n\pi\frac{z'_D}{z_{eD}}$$
$$\left[\frac{ch\sqrt{u+\frac{n^2\pi^2}{z_{eD}^2}}(y_{eD}-|\tilde{y}_{D1}|)+ch\sqrt{u+\frac{n^2\pi^2}{z_{eD}^2}}(y_{eD}-\tilde{y}_{D2})}{\sqrt{u+\frac{n^2\pi^2}{z_{eD}^2}}sh\sqrt{u+\frac{n^2\pi^2}{z_{eD}^2}}y_{eD}}\right.$$
$$+2\sum_{k=1}^{\infty}\cos k\pi\frac{x_D}{x_{eD}}\cos k\pi\frac{x'_D}{x_{eD}}$$
$$\left.\left.\frac{ch\sqrt{u+\frac{k^2\pi^2}{x_{eD}^2}+\frac{n^2\pi^2}{z_{eD}^2}}(y_{eD}-|\tilde{y}_{D1}|)+ch\sqrt{u+\frac{k^2\pi^2}{x_{eD}^2}+\frac{n^2\pi^2}{z_{eD}^2}}(y_{eD}-\tilde{y}_{D2})}{\sqrt{u+\frac{k^2\pi^2}{x_{eD}^2}+\frac{n^2\pi^2}{z_{eD}^2}}sh\sqrt{u+\frac{k^2\pi^2}{x_{eD}^2}+\frac{n^2\pi^2}{z_{eD}^2}}y_{eD}}\right]\right\}. \tag{13}$$

II. *One boundary is at the initial pressure, the others are impermeable*

Let the surface $x_D = x_{eD}$ be at the initial pressure and all other surfaces be impermeable. The fundamental solution for this case is given by

$$\overline{\gamma} = \frac{1}{4\pi} \sum_{k=-\infty}^{+\infty} \sum_{m=-\infty}^{+\infty} \sum_{n=-\infty}^{+\infty} (-1)^k \\ (S_{1,1,1} + S_{2,1,1} + S_{1,2,1} + S_{2,2,1} + S_{1,1,2} + S_{2,1,2} + S_{1,2,2} + S_{2,2,2}). \tag{14}$$

If we let

$$T_S = \sum_{k=-\infty}^{+\infty} \sum_{m=-\infty}^{+\infty} \sum_{n=-\infty}^{+\infty} (-1)^k S, \tag{15}$$

where S is given by (5), then, using 2.2(5) and (6), we have

$$T_S = \frac{1}{2z_{eD}} \left\{ \int_0^{\infty} \exp(-\xi) \sum_{k=-\infty}^{+\infty} (-1)^k \exp\left[-\frac{u(\widetilde{x}_D - 2kx_{eD})^2}{4\xi} \right] \right. \\ \sum_{m=-\infty}^{+\infty} \exp\left[-\frac{u(\widetilde{y}_D - 2my_{eD})^2}{4\xi} \right] \frac{d\xi}{\xi} \\ + 2\sum_{n=1}^{\infty} \cos n\pi \frac{\widetilde{z}_D}{z_{eD}} \int_0^{\infty} \exp(-\xi) \sum_{k=-\infty}^{+\infty} (-1)^k \exp\left[-\frac{(u+a)(\widetilde{x}_D - 2kx_{eD})^2}{4\xi} \right] \\ \left. \sum_{m=-\infty}^{+\infty} \exp\left[-\frac{(u+a)(\widetilde{y}_D - 2my_{eD})^2}{4\xi} \right] \frac{d\xi}{\xi} \right\}. \tag{16}$$

Here, a is given by $n^2\pi^2/z_{eD}^2$. If we now let T_1 represent the left-hand side of 2.2(11), then using Poisson's summation formula, 2.2(5), we obtain

$$T_1 = \sum_{k=-\infty}^{+\infty} (-1)^k \exp\left[-\frac{a(\widetilde{x}_D - 2kx_{eD})^2}{4\xi} \right] \\ = \frac{2\sqrt{\pi\xi}}{\sqrt{a}x_{eD}} \sum_{k=1}^{\infty} \exp\left[-\frac{(2k-1)^2\pi^2\xi}{4ax_{eD}^2} \right] \cos(2k-1)\frac{\pi}{2}\frac{\widetilde{x}_D}{x_{eD}}. \tag{17}$$

Using (17), we may write (16) as follows

$$
\begin{aligned}
T_S = \frac{\pi}{x_{eD} y_{eD} z_{eD}} \Bigg\{ & \frac{1}{u} \sum_{k=1}^{\infty} \frac{\cos(2k-1)\frac{\pi}{2}\frac{\widetilde{x}_D}{x_{eD}}}{1 + \frac{(2k-1)^2\pi^2}{4u x_{eD}^2}} \\
& + \frac{2}{u} \sum_{k=1}^{\infty} \cos(2k-1)\frac{\pi}{2}\frac{\widetilde{x}_D}{x_{eD}} \sum_{m=1}^{\infty} \frac{\cos m\pi \frac{\widetilde{y}_D}{y_{eD}}}{1 + \frac{(2k-1)^2\pi^2}{4u x_{eD}^2} + \frac{m^2\pi^2}{u y_{eD}^2}} \\
& + 2 \sum_{n=1}^{\infty} \frac{\cos n\pi \frac{\widetilde{z}_D}{z_{eD}}}{u+a} \sum_{k=1}^{\infty} \frac{\cos(2k-1)\frac{\pi}{2}\frac{\widetilde{x}_D}{x_{eD}}}{1 + \frac{(2k-1)^2\pi^2}{4(u+a) x_{eD}^2}} \\
& + 4 \sum_{n=1}^{\infty} \frac{\cos n\pi \frac{\widetilde{z}_D}{z_{eD}}}{u+a} \sum_{k=1}^{\infty} \cos(2k-1)\frac{\pi}{2}\frac{\widetilde{x}_D}{x_{eD}} \sum_{m=1}^{\infty} \frac{\cos m\pi \frac{\widetilde{y}_D}{y_{eD}}}{1 + \frac{(2k-1)^2\pi^2}{4(u+a) x_{eD}^2} + \frac{m^2\pi^2}{(u+a) y_{eD}^2}} \Bigg\}.
\end{aligned}
\tag{18}
$$

Using (10), we have

$$
\begin{aligned}
T_S = \frac{\pi}{x_{eD} z_{eD}} \Bigg\{ & \sum_{k=1}^{\infty} \frac{\cos(2k-1)\frac{\pi}{2}\frac{\widetilde{x}_D}{x_{eD}}}{\sqrt{u + \frac{(2k-1)^2\pi^2}{4x_{eD}^2}}} \frac{ch\sqrt{u + \frac{(2k-1)^2\pi^2}{4x_{eD}^2}}(y_{eD} - |\widetilde{y}_D|)}{sh\sqrt{u + \frac{(2k-1)^2\pi^2}{4x_{eD}^2}}\, y_{eD}} \\
& + 2 \sum_{n=1}^{\infty} \cos n\pi \frac{\widetilde{z}_D}{z_{eD}} \sum_{k=1}^{\infty} \frac{\cos(2k-1)\frac{\pi}{2}\frac{\widetilde{x}_D}{x_{eD}}}{\sqrt{u + \frac{(2k-1)^2\pi^2}{4x_{eD}^2} + \frac{n^2\pi^2}{z_{eD}^2}}} \\
& \frac{ch\sqrt{u + \frac{(2k-1)^2\pi^2}{4x_{eD}^2} + \frac{n^2\pi^2}{z_{eD}^2}}(y_{eD} - |\widetilde{y}_D|)}{sh\sqrt{u + \frac{(2k-1)^2\pi^2}{4x_{eD}^2} + \frac{n^2\pi^2}{z_{eD}^2}}\, y_{eD}} \Bigg\}.
\end{aligned}
\tag{19}
$$

Thus, we can write the fundamental solution as

$$
\begin{aligned}
\overline{\gamma} = \frac{1}{x_{eD} z_{eD}} \Bigg\{ & \sum_{k=1}^{\infty} \cos(2k-1)\frac{\pi}{2}\frac{x_D}{x_{eD}} \cos(2k-1)\frac{\pi}{2}\frac{x'_D}{x_{eD}} \\
& \left[\frac{ch\sqrt{u + \frac{(2k-1)^2\pi^2}{4x_{eD}^2}}(y_{eD} - |\widetilde{y}_{D1}|) + ch\sqrt{u + \frac{(2k-1)^2\pi^2}{4x_{eD}^2}}(y_{eD} - \widetilde{y}_{D2})}{\sqrt{u + \frac{(2k-1)^2\pi^2}{4x_{eD}^2}}\, sh\sqrt{u + \frac{(2k-1)^2\pi^2}{4x_{eD}^2}}\, y_{eD}} \right. \\
& + 2 \sum_{n=1}^{\infty} \cos n\pi \frac{z_D}{z_{eD}} \cos n\pi \frac{z'_D}{z_{eD}} \\
& \left. \frac{ch\sqrt{u + \frac{n^2\pi^2}{z_{eD}^2} + \frac{(2k-1)^2\pi^2}{4x_{eD}^2}}(y_{eD} - |\widetilde{y}_{D1}|) + ch\sqrt{u + \frac{n^2\pi^2}{z_{eD}^2} + \frac{(2k-1)^2\pi^2}{4x_{eD}^2}}(y_{eD} - \widetilde{y}_{D2})}{\sqrt{u + \frac{n^2\pi^2}{z_{eD}^2} + \frac{(2k-1)^2\pi^2}{4x_{eD}^2}}\, sh\sqrt{u + \frac{n^2\pi^2}{z_{eD}^2} + \frac{(2k-1)^2\pi^2}{4x_{eD}^2}}\, y_{eD}} \right] \Bigg\}.
\end{aligned}
\tag{20}
$$

III. *Two boundaries are at the initial pressure, the others are impermeable*

We consider two possible combinations. In the first case, the constant-pressure boundaries are opposite to each other and in the second case the boundaries are adjacent to each other.

A. The boundaries $y_D = 0$ and $y_D = y_{eD}$ are at the initial pressure.

The fundamental solution in this case is

$$\overline{\gamma} = \frac{1}{4\pi} \sum_{k=-\infty}^{+\infty} \sum_{m=-\infty}^{+\infty} \sum_{n=-\infty}^{+\infty} (S_{1,1,1} + S_{2,1,1} - S_{1,2,1} - S_{2,2,1} + S_{1,1,2} + S_{2,1,2} - S_{1,2,2} - S_{2,2,2}). \tag{21}$$

The appropriate expression for $\overline{\gamma}$ can be obtained by using (12) for each summation term in (21). Simplification yields

$$\overline{\gamma} = \frac{1}{2x_{eD}z_{eD}} \left\{ \frac{ch\sqrt{u}(y_{eD} - |\widetilde{y}_{D1}|) - ch\sqrt{u}(y_{eD} - \widetilde{y}_{D2})}{\sqrt{u}sh\sqrt{u}y_{eD}} + 2\sum_{k=1}^{\infty} \cos k\pi\frac{x_D}{x_{eD}} \cos k\pi\frac{x'_D}{x_{eD}} \right.$$
$$\frac{ch\sqrt{u + \frac{k^2\pi^2}{x_{eD}^2}}(y_{eD} - |\widetilde{y}_{D1}|) - ch\sqrt{u + \frac{k^2\pi^2}{x_{eD}^2}}(y_{eD} - \widetilde{y}_{D2})}{\sqrt{u + \frac{k^2\pi^2}{x_{eD}^2}}sh\sqrt{u + \frac{k^2\pi^2}{x_{eD}^2}}y_{eD}}$$
$$+ 2\sum_{n=1}^{\infty} \cos n\pi\frac{z_D}{z_{eD}} \cos n\pi\frac{z'_D}{z_{eD}}$$
$$\left[\frac{ch\sqrt{u + \frac{n^2\pi^2}{z_{eD}^2}}(y_{eD} - |\widetilde{y}_{D1}|) - ch\sqrt{u + \frac{n^2\pi^2}{z_{eD}^2}}(y_{eD} - \widetilde{y}_{D2})}{\sqrt{u + \frac{n^2\pi^2}{z_{eD}^2}}sh\sqrt{u + \frac{n^2\pi^2}{z_{eD}^2}}y_{eD}} \right.$$
$$+ 2\sum_{k=1}^{\infty} \cos k\pi\frac{x_D}{x_{eD}} \cos k\pi\frac{x'_D}{x_{eD}}$$
$$\left. \left. \frac{ch\sqrt{u + \frac{k^2\pi^2}{x_{eD}^2} + \frac{n^2\pi^2}{z_{eD}^2}}(y_{eD} - |\widetilde{y}_{D1}|) - ch\sqrt{u + \frac{k^2\pi^2}{x_{eD}^2} + \frac{n^2\pi^2}{z_{eD}^2}}(y_{eD} - \widetilde{y}_{D2})}{\sqrt{u + \frac{k^2\pi^2}{x_{eD}^2} + \frac{n^2\pi^2}{z_{eD}^2}}sh\sqrt{u + \frac{k^2\pi^2}{x_{eD}^2} + \frac{n^2\pi^2}{z_{eD}^2}}y_{eD}} \right] \right\}. \tag{22}$$

B. The boundaries $x_D = x_{eD}$ and $y_D = y_{eD}$ are at the initial pressure.

The fundamental solution in this case is

$$\overline{\gamma} = \frac{1}{4\pi} \sum_{k=-\infty}^{+\infty} \sum_{m=-\infty}^{+\infty} \sum_{n=-\infty}^{+\infty} (-1)^k(-1)^m (S_{1,1,1} + S_{2,1,1} + S_{1,2,1} + S_{2,2,1} + S_{1,1,2} + S_{2,1,2} + S_{1,2,2} + S_{2,2,2}). \tag{23}$$

If we let T_S represent one of the terms in (23), we have, using 2.2(6) and (6),

$$T_S = \frac{1}{2z_{eD}} \left\{ \int_0^\infty \exp(-\xi) \sum_{k=-\infty}^{+\infty} (-1)^k \exp\left[-\frac{u(\widetilde{x}_D - 2kx_{eD})^2}{4\xi}\right] \right.$$
$$\sum_{m=-\infty}^{+\infty} (-1)^m \exp\left[-\frac{u(\widetilde{y}_D - 2my_{eD})^2}{4\xi}\right] \frac{d\xi}{\xi}$$
$$+ 2\sum_{n=1}^{\infty} \cos n\pi \frac{\widetilde{z}_D}{z_{eD}} \int_0^\infty \exp(-\xi) \sum_{k=-\infty}^{+\infty} (-1)^k \exp\left[-\frac{(u+a)(\widetilde{x}_D - 2kx_{eD})^2}{4\xi}\right] \tag{24}$$
$$\left. \sum_{m=-\infty}^{+\infty} (-1)^m \exp\left[-\frac{(u+a)(\widetilde{y}_D - 2my_{eD})^2}{4\xi}\right] \frac{d\xi}{\xi} \right\},$$

where $a = n^2\pi^2/z_{eD}^2$. Using the result given in (17),

$$T_S = \frac{2\pi}{x_{eD}y_{eD}z_{eD}u} \sum_{k=1}^{\infty} \cos(2k-1)\frac{\pi}{2}\frac{\widetilde{x}_D}{x_{eD}} \sum_{m=1}^{\infty} \frac{\cos(2m-1)\frac{\pi}{2}\frac{\widetilde{y}_D}{y_{eD}}}{1 + \frac{(2k-1)^2\pi^2}{4ux_{eD}^2} + \frac{(2m-1)^2\pi^2}{4uy_{eD}^2}}$$
$$+ \frac{4\pi}{x_{eD}y_{eD}z_{eD}} \sum_{n=1}^{\infty} \frac{\cos n\pi \frac{\widetilde{z}_D}{z_{eD}}}{u+a} \sum_{k=1}^{\infty} \cos(2k-1)\frac{\pi}{2}\frac{\widetilde{x}_D}{x_{eD}} \tag{25}$$
$$\sum_{m=1}^{\infty} \frac{\cos(2m-1)\frac{\pi}{2}\frac{\widetilde{y}_D}{y_{eD}}}{1 + \frac{(2k-1)^2\pi^2}{4\left(u+\frac{n^2\pi^2}{z_{eD}^2}\right)x_{eD}^2} + \frac{(2m-1)^2\pi^2}{4\left(u+\frac{n^2\pi^2}{z_{eD}^2}\right)y_{eD}^2}}.$$

Because (see Hansen [8], Eq. 17.3.12)

$$\sum_{k=1}^{\infty} \frac{\cos(2k-1)\frac{\pi}{2}x}{b^2 + (2k-1)^2} = \frac{\pi}{4b}\frac{sh\frac{\pi b}{2}(1-x)}{ch\frac{\pi b}{2}}; \quad 0 \le x \le 2, \tag{26}$$

$$T_S = \frac{\pi}{x_{eD}z_{eD}} \left\{ \sum_{k=1}^{\infty} \cos(2k-1)\frac{\pi}{2}\frac{\widetilde{x}_D}{x_{eD}} \frac{sh\sqrt{u + \frac{(2k-1)^2\pi^2}{4x_{eD}^2}}(y_{eD} - |\widetilde{y}_D|)}{\sqrt{u + \frac{(2k-1)^2\pi^2}{4x_{eD}^2}}\, ch\sqrt{u + \frac{(2k-1)^2\pi^2}{4x_{eD}^2}}\, y_{eD}} \right.$$
$$+ 2\sum_{n=1}^{\infty} \cos n\pi \frac{\widetilde{z}_D}{z_{eD}} \tag{27}$$
$$\left. \sum_{k=1}^{\infty} \frac{\cos(2k-1)\frac{\pi}{2}\frac{\widetilde{x}_D}{x_{eD}}}{\sqrt{u + \frac{(2k-1)^2\pi^2}{4x_{eD}^2} + \frac{n^2\pi^2}{z_{eD}^2}}} \frac{sh\sqrt{u + \frac{(2k-1)^2\pi^2}{4x_{eD}^2} + \frac{n^2\pi^2}{z_{eD}^2}}(y_{eD} - |\widetilde{y}_D|)}{ch\sqrt{u + \frac{(2k-1)^2\pi^2}{4x_{eD}^2} + \frac{n^2\pi^2}{z_{eD}^2}}\, y_{eD}} \right\}.$$

Thus,

$$\overline{\gamma} = \frac{1}{x_{eD}z_{eD}}\left\{\sum_{k=1}^{\infty} \cos(2k-1)\frac{\pi}{2}\frac{x_D}{x_{eD}} \cos(2k-1)\frac{\pi}{2}\frac{x'_D}{x_{eD}}\right.$$

$$\left[\frac{sh\sqrt{u+\frac{(2k-1)^2\pi^2}{4x_{eD}^2}}(y_{eD}-|\widetilde{y}_{D1}|)+sh\sqrt{u+\frac{(2k-1)^2\pi^2}{4x_{eD}^2}}(y_{eD}-\widetilde{y}_{D2})}{\sqrt{u+\frac{(2k-1)^2\pi^2}{4x_{eD}^2}}ch\sqrt{u+\frac{(2k-1)^2\pi^2}{4x_{eD}^2}}y_{eD}}\right.$$

$$+2\sum_{n=1}^{\infty}\cos n\pi\frac{z_D}{z_{eD}}\cos n\pi\frac{z'_D}{z_{eD}}$$

$$\left.\left.\frac{sh\sqrt{u+\frac{(2k-1)^2\pi^2}{4x_{eD}^2}+\frac{n^2\pi^2}{z_{eD}^2}}(y_{eD}-|\widetilde{y}_{D1}|)+sh\sqrt{u+\frac{(2k-1)^2\pi^2}{4x_{eD}^2}+\frac{n^2\pi^2}{z_{eD}^2}}(y_{eD}-\widetilde{y}_{D2})}{\sqrt{u+\frac{(2k-1)^2\pi^2}{4x_{eD}^2}+\frac{n^2\pi^2}{z_{eD}^2}}ch\sqrt{u+\frac{(2k-1)^2\pi^2}{4x_{eD}^2}+\frac{n^2\pi^2}{z_{eD}^2}}y_{eD}}\right]\right\}. \tag{28}$$

IV. *Three boundaries are at the initial pressure, the others are impermeable*

Again we need to consider two cases.

A. The boundaries $x_D = x_{eD}$, $y_D = y_{eD}$, and $z_D = z_{eD}$ are at the initial pressure.

The fundamental solution in this case is

$$\overline{\gamma} = \frac{1}{4\pi}\sum_{k=-\infty}^{+\infty}\sum_{m=-\infty}^{+\infty}\sum_{n=-\infty}^{+\infty}(-1)^k(-1)^m(-1)^n$$
$$(S_{1,1,1}+S_{2,1,1}+S_{1,2,1}+S_{2,2,1}+S_{1,1,2}+S_{2,1,2}+S_{1,2,2}+S_{2,2,2}). \tag{29}$$

We now proceed to evaluate one of the terms in (29). Let T_S represent this term. We first write, using 2.2(6) and 2.2(11),

$$T_S = \sum_{k=-\infty}^{+\infty}\sum_{m=-\infty}^{+\infty}\sum_{n=-\infty}^{+\infty}(-1)^k(-1)^m(-1)^n$$
$$\frac{\exp\left[-\sqrt{u}\sqrt{(\widetilde{x}_D-2kx_{eD})^2+(\widetilde{y}_D-2my_{eD})^2+(\widetilde{z}_D-2nz_{eD})^2}\right]}{\sqrt{(\widetilde{x}_D-2kx_{eD})^2+(\widetilde{y}_D-2my_{eD})^2+(\widetilde{z}_D-2nz_{eD})^2}}$$
$$=\frac{2}{z_{eD}}\sum_{k=-\infty}^{+\infty}\sum_{m=-\infty}^{+\infty}(-1)^k(-1)^m \tag{30}$$
$$\sum_{n=1}^{\infty}K_0\left[\sqrt{u+a}\sqrt{(\widetilde{x}_D-2kx_{eD})^2+(\widetilde{y}_D-2my_{eD})^2}\right]\cos(2n-1)\frac{\pi}{2}\frac{\widetilde{z}_D}{z_{eD}},$$

where $a = \frac{(2n-1)^2\pi^2}{4z_{eD}^2}$. Using (6), we have

$$T_S = \frac{1}{z_{eD}} \sum_{n=1}^{\infty} \left\{ \int_0^{\infty} \exp(-\xi) \sum_{k=-\infty}^{+\infty} (-1)^k \exp\left[-\frac{(u+a)(\widetilde{x}_D - 2kx_{eD})^2}{4\xi} \right] \right. \\ \left. \sum_{m=-\infty}^{+\infty} (-1)^m \exp\left[-\frac{(u+a)(\widetilde{y}_D - 2my_{eD})^2}{4\xi} \right] \frac{d\xi}{\xi} \right\} \cos(2n-1)\frac{\pi}{2}\frac{\widetilde{z}_D}{z_{eD}}. \tag{31}$$

Using Poisson's summation formula, 2.2(5) and 2.2(11), we can write (31) as

$$T_S = \frac{4\pi}{x_{eD}y_{eD}z_{eD}} \sum_{n=1}^{\infty} \frac{\cos(2n-1)\frac{\pi}{2}\frac{\widetilde{z}_D}{z_{eD}}}{(u+a)} \sum_{k=1}^{\infty} \cos(2k-1)\frac{\pi}{2}\frac{\widetilde{x}_D}{x_{eD}} \sum_{m=1}^{\infty} \cos(2m-1)\frac{\pi}{2}\frac{\widetilde{y}_D}{y_{eD}} \\ \int_0^{\infty} \exp(-\xi)\exp(-b\xi)d\xi, \tag{32}$$

where

$$b = \frac{(2k-1)^2\pi^2}{4(u+a)x_{eD}^2} + \frac{(2m-1)^2\pi^2}{4(u+a)y_{eD}^2}. \tag{33}$$

Evaluating the integral in (32), and using (26), we have

$$T_S = \frac{2\pi}{x_{eD}z_{eD}} \left[\sum_{n=1}^{\infty} \cos(2n-1)\frac{\pi}{2}\frac{\widetilde{z}_D}{z_{eD}} \sum_{k=1}^{\infty} \cos(2k-1)\frac{\pi}{2}\frac{\widetilde{x}_D}{x_{eD}} \right. \\ \left. \frac{sh\sqrt{u + \frac{(2k-1)^2\pi^2}{4x_{eD}^2} + \frac{(2n-1)^2\pi^2}{4z_{eD}^2}}(y_{eD} - |\widetilde{y}_D|)}{\sqrt{u + \frac{(2k-1)^2\pi^2}{4x_{eD}^2} + \frac{(2n-1)^2\pi^2}{4z_{eD}^2}}\, ch\sqrt{u + \frac{(2k-1)^2\pi^2}{4x_{eD}^2} + \frac{(2n-1)^2\pi^2}{4z_{eD}^2}}\, y_{eD}} \right]. \tag{34}$$

Thus, using (34), the fundamental solution given in (29) is given by

$$\overline{\gamma} = \frac{2}{x_{eD}z_{eD}} \sum_{n=1}^{\infty} \cos(2n-1)\frac{\pi}{2}\frac{z_D}{z_{eD}} \cos(2n-1)\frac{\pi}{2}\frac{z'_D}{z_{eD}} \\ \sum_{k=1}^{\infty} \cos(2k-1)\frac{\pi}{2}\frac{x_D}{x_{eD}} \cos(2k-1)\frac{\pi}{2}\frac{x'_D}{x_{eD}} \\ \frac{sh\sqrt{u + \frac{(2k-1)^2\pi^2}{4x_{eD}^2} + \frac{(2n-1)^2\pi^2}{4z_{eD}^2}}(y_{eD} - |\widetilde{y}_{D1}|) + sh\sqrt{u + \frac{(2k-1)^2\pi^2}{4x_{eD}^2} + \frac{(2n-1)^2\pi^2}{4z_{eD}^2}}(y_{eD} - \widetilde{y}_{D2})}{\sqrt{u + \frac{(2k-1)^2\pi^2}{4x_{eD}^2} + \frac{(2n-1)^2\pi^2}{4z_{eD}^2}}\, ch\sqrt{u + \frac{(2k-1)^2\pi^2}{4x_{eD}^2} + \frac{(2n-1)^2\pi^2}{4z_{eD}^2}}\, y_{eD}}. \tag{35}$$

B. The boundaries $x_D = x_{eD}$, $y_D = 0$ and $y_D = y_{eD}$ are at the initial pressure.

The fundamental solution in this case is given by

$$\overline{\gamma} = \frac{1}{4\pi} \sum_{k=-\infty}^{+\infty} \sum_{m=-\infty}^{+\infty} \sum_{n=-\infty}^{+\infty} (-1)^k (S_{1,1,1} + S_{2,1,1} - S_{1,2,1} - S_{2,2,1} \\ + S_{1,1,2} + S_{2,1,2} - S_{1,2,2} - S_{2,2,2}). \tag{36}$$

We have already developed the appropriate expressions in II. Using the expression given in (19), we may write

$$\overline{\gamma} = \frac{1}{x_{eD} z_{eD}} \left\{ \sum_{k=1}^{\infty} \cos(2k-1)\frac{\pi}{2}\frac{x_D}{x_{eD}} \cos(2k-1)\frac{\pi}{2}\frac{x'_D}{x_{eD}} \right.$$

$$\left[\frac{ch\sqrt{u+\frac{(2k-1)^2\pi^2}{4x_{eD}^2}}(y_{eD}-|\widetilde{y}_{D1}|) - ch\sqrt{u+\frac{(2k-1)^2\pi^2}{4x_{eD}^2}}(y_{eD}-\widetilde{y}_{D2})}{\sqrt{u+\frac{(2k-1)^2\pi^2}{4x_{eD}^2}}\, sh\sqrt{u+\frac{(2k-1)^2\pi^2}{4x_{eD}^2}}\,\widetilde{y}_{eD}} \right.$$

$$+ 2\sum_{n=1}^{\infty} \cos n\pi\frac{z_D}{z_{eD}} \cos n\pi\frac{z'_D}{z_{eD}}$$

$$\left.\left. \frac{ch\sqrt{u+\frac{(2k-1)^2\pi^2}{4x_{eD}^2}+\frac{n^2\pi^2}{z_{eD}^2}}(y_{eD}-|\widetilde{y}_{D1}|) - ch\sqrt{u+\frac{(2k-1)^2\pi^2}{4x_{eD}^2}+\frac{n^2\pi^2}{z_{eD}^2}}(y_{eD}-\widetilde{y}_{D2})}{\sqrt{u+\frac{(2k-1)^2\pi^2}{4x_{eD}^2}+\frac{n^2\pi^2}{z_{eD}^2}}\, sh\sqrt{u+\frac{(2k-1)^2\pi^2}{4x_{eD}^2}+\frac{n^2\pi^2}{z_{eD}^2}}\, y_{eD}} \right] \right\}. \tag{37}$$

V. *Four boundaries are at the initial pressure, the others are impermeable*

We consider two cases.

A. The boundaries $x_D = x_{eD}$, $y_D = y_{eD}$, $z_D = 0$, and $z_D = z_{eD}$ are at the initial pressure.

The fundamental solution by the method of images is

$$\overline{\gamma} = \frac{1}{4\pi} \sum_{k=-\infty}^{+\infty} \sum_{m=-\infty}^{+\infty} \sum_{n=-\infty}^{+\infty} (-1)^k(-1)^m$$
$$(S_{1,1,1} + S_{2,1,1} + S_{1,2,1} + S_{2,2,1} - S_{1,1,2} - S_{2,1,2} - S_{1,2,2} - S_{2,2,2}). \tag{38}$$

We outlined in III the steps to evaluate

$$T_S = \sum_{k=-\infty}^{+\infty} \sum_{m=-\infty}^{+\infty} \sum_{n=-\infty}^{+\infty} (-1)^k(-1)^m S. \tag{39}$$

Using the result in (27), the fundamental solution for this system is

$$\overline{\gamma} = \frac{2}{x_{eD} z_{eD}} \left[\sum_{n=1}^{\infty} \sin n\pi\frac{z_D}{z_{eD}} \sin n\pi\frac{z'_D}{z_{eD}} \right.$$

$$\sum_{k=1}^{\infty} \cos(2k-1)\frac{\pi}{2}\frac{x_D}{x_{eD}} \cos(2k-1)\frac{\pi}{2}\frac{x'_D}{x_{eD}}$$

$$\left. \frac{sh\sqrt{u+\frac{(2k-1)^2\pi^2}{4x_{eD}^2}+\frac{n^2\pi^2}{z_{eD}^2}}(y_{eD}-|\widetilde{y}_{D1}|) + sh\sqrt{u+\frac{(2k-1)^2\pi^2}{4x_{eD}^2}+\frac{n^2\pi^2}{z_{eD}^2}}(y_{eD}-\widetilde{y}_{D2})}{\sqrt{u+\frac{(2k-1)^2\pi^2}{4x_{eD}^2}+\frac{n^2\pi^2}{z_{eD}^2}}\, ch\sqrt{u+\frac{(2k-1)^2\pi^2}{4x_{eD}^2}+\frac{n^2\pi^2}{z_{eD}^2}}\, y_{eD}} \right]. \tag{40}$$

B. The boundaries $x_D = 0$, $x_D = x_{eD}$, $y_D = 0$, and $y_D = y_{eD}$ are at the initial pressure.

The fundamental solution by the method of images is

$$\overline{\gamma} = \frac{1}{4\pi} \sum_{k=-\infty}^{+\infty} \sum_{m=-\infty}^{+\infty} \sum_{n=-\infty}^{+\infty} \left(S_{1,1,1} - S_{2,1,1} - S_{1,2,1} + S_{2,2,1} + S_{1,1,2} - S_{2,1,2} - S_{1,2,2} + S_{2,2,2}\right). \tag{41}$$

We can use the expression in (12) to evaluate the triple sums in (41). Thus, the fundamental solution is given by

$$\begin{aligned}\overline{\gamma} = \frac{1}{x_{eD} z_{eD}} \Bigg[&\sum_{k=1}^{\infty} \sin k\pi \frac{x_D}{x_{eD}} \sin k\pi \frac{x'_D}{x_{eD}} \\ &\frac{ch\sqrt{u + \frac{k^2\pi^2}{x_{eD}^2}}(y_{eD} - |\widetilde{y}_{D1}|) - ch\sqrt{u + \frac{k^2\pi^2}{x_{eD}^2}}(y_{eD} - \widetilde{y}_{D2})}{\sqrt{u + \frac{k^2\pi^2}{x_{eD}^2}}\, sh\sqrt{u + \frac{k^2\pi^2}{x_{eD}^2}}\, y_{eD}} \\ &+ 2\sum_{n=1}^{\infty} \cos n\pi \frac{z_D}{z_{eD}} \cos n\pi \frac{z'_D}{z_{eD}} \sum_{k=1}^{\infty} \sin k\pi \frac{x_D}{x_{eD}} \sin k\pi \frac{x'_D}{x_{eD}} \\ &\frac{ch\sqrt{u + \frac{k^2\pi^2}{x_{eD}^2} + \frac{n^2\pi^2}{z_{eD}^2}}(y_{eD} - |\widetilde{y}_{D1}|) - ch\sqrt{u + \frac{k^2\pi^2}{x_{eD}^2} + \frac{n^2\pi^2}{z_{eD}^2}}(y_{eD} - \widetilde{y}_{D2})}{\sqrt{u + \frac{k^2\pi^2}{x_{eD}^2} + \frac{n^2\pi^2}{z_{eD}^2}}\, sh\sqrt{u + \frac{k^2\pi^2}{x_{eD}^2} + \frac{n^2\pi^2}{z_{eD}^2}}\, y_{eD}} \Bigg].\end{aligned} \tag{42}$$

VI. *Five boundaries are at the initial pressure, the other is impermeable*

Let the surface $x_D = 0$ be impermeable and the other boundaries be at a the initial pressure. The fundamental solution in this case is given by

$$\begin{aligned}\overline{\gamma} = \frac{1}{4\pi} \sum_{k=-\infty}^{+\infty} \sum_{m=-\infty}^{+\infty} \sum_{n=-\infty}^{+\infty} (-1)^k &\left(S_{1,1,1} + S_{2,1,1} - S_{1,2,1} - S_{2,2,1}\right. \\ &\left. - S_{1,1,2} - S_{2,1,2} + S_{1,2,2} + S_{2,2,2}\right).\end{aligned} \tag{43}$$

Using the results in II (see (19)), we have

$$\begin{aligned}\overline{\gamma} = \frac{1}{x_{eD} z_{eD}} \Bigg\{ &\sum_{k=1}^{\infty} \cos(2k-1)\frac{\pi}{2}\frac{x_D}{x_{eD}} \cos(2k-1)\frac{\pi}{2}\frac{x'_D}{x_{eD}} \\ &\Bigg[\frac{ch\sqrt{u + \frac{(2k-1)^2\pi^2}{4x_{eD}^2}}(y_{eD} - |\widetilde{y}_{D1}|) - ch\sqrt{u + \frac{(2k-1)^2\pi^2}{4x_{eD}^2}}(y_{eD} - \widetilde{y}_{D2})}{\sqrt{u + \frac{(2k-1)^2\pi^2}{4x_{eD}^2}}\, sh\sqrt{u + \frac{(2k-1)^2\pi^2}{4x_{eD}^2}}\, y_{eD}} \\ &+ 2\sum_{n=1}^{\infty} \sin n\pi \frac{z_D}{z_{eD}} \sin n\pi \frac{z'_D}{z_{eD}} \\ &\frac{ch\sqrt{u + \frac{(2k-1)^2\pi^2}{4x_{eD}^2} + \frac{n^2\pi^2}{z_{eD}^2}}(y_{eD} - |\widetilde{y}_{D1}|) - ch\sqrt{u + \frac{(2k-1)^2\pi^2}{4x_{eD}^2} + \frac{n^2\pi^2}{z_{eD}^2}}(y_{eD} - \widetilde{y}_{D2})}{\sqrt{u + \frac{(2k-1)^2\pi^2}{4x_{eD}^2} + \frac{n^2\pi^2}{z_{eD}^2}}\, sh\sqrt{u + \frac{(2k-1)^2\pi^2}{4x_{eD}^2} + \frac{n^2\pi^2}{z_{eD}^2}}\, y_{eD}} \Bigg]\Bigg\}.\end{aligned} \tag{44}$$

VII. *All boundaries are at the initial pressure*

Using the method of images, the fundamental solution is given by

$$\overline{\gamma} = \frac{1}{4\pi} \sum_{k=-\infty}^{+\infty} \sum_{m=-\infty}^{+\infty} \sum_{n=-\infty}^{+\infty} \left(S_{1,1,1} - S_{2,1,1} - S_{1,2,1} + S_{2,2,1} - S_{1,1,2} + S_{2,1,2} + S_{1,2,2} - S_{2,2,2}\right). \tag{45}$$

Using the expression for T_S derived in (12), we have

$$\overline{\gamma} = \frac{2}{x_{eD} z_{eD}} \left\{ \sum_{n=1}^{\infty} \sin n\pi \frac{z_D}{z_{eD}} \sin n\pi \frac{z'_D}{z_{eD}} \sum_{k=1}^{\infty} \sin k\pi \frac{x_D}{x_{eD}} \sin k\pi \frac{x'_D}{x_{eD}} \right. \\ \left. \frac{ch\sqrt{u + \frac{k^2\pi^2}{x_{eD}^2} + \frac{n^2\pi^2}{z_{eD}^2}}(y_{eD} - |\widetilde{y}_{D1}|) - ch\sqrt{u + \frac{k^2\pi^2}{x_{eD}^2} + \frac{n^2\pi^2}{z_{eD}^2}}(y_{eD} - \widetilde{y}_{D2})}{\sqrt{u + \frac{k^2\pi^2}{x_{eD}^2} + \frac{n^2\pi^2}{z_{eD}^2}}\, sh\sqrt{u + \frac{k^2\pi^2}{x_{eD}^2} + \frac{n^2\pi^2}{z_{eD}^2}}\, y_{eD}} \right\}. \tag{46}$$

This completes our discussion of the solutions needed to solve for pressure distributions in porous solids that are considered to be rectangular parallelepipeds.

References

1. Carslaw, H. S. and Jaeger, J. C.: *Conduction of Heat in Solids*, Oxford University Press, Oxford (1959), 275.
2. Ozkan, E.: *Performance of Horizontal Wells*, Ph. D. Dissertation, The University of Tulsa, Tulsa, OK (1988), 17.
3. Carslaw, H. S. and Jaeger, J. C.: *op. cit.*, 353–386.
4. Muskat, M.: *The Flow of Homogeneous Fluids Through Porous Media*, McGraw-Hill (1937), 263–277.
5. Carslaw, H. S. and Jaeger, J. C.: *op. cit.*, 377.
6. Watson, G. N.: *A Treatise on the Theory of Bessel Functions*, Cambridge University Press, London (1948), 80.
7. Gradshteyn, I. S. and Ryzhik, I. M.: *Table of Integrals, Series, and Products*, Academic Press, Inc., Orlando (1980), 40.
8. Hansen, E. R.: *A Table of Series and Products*, Prentice–Hall, Inc., Englewood Cliffs, New Jersey (1975), 243.

III. Treatment of Extraction Points

A unique feature of the production of fluids from geologic media is that for a variety of reasons, the surfaces through which fluid has to be extracted have complicated geometrical features. The conventional wellbore (the simplest case) is assumed to be a cylindrical surface that extends over the entire thickness of the porous medium (in all of the following we assume that the porous medium is horizontal). Such a wellbore is usually approximated by a line source. When extracting fluids from geologic media, it is not unusual for wellbores to be in contact with only a part of the porous medium; that is, fluid is not extracted over the entire thickness of the porous medium, and in this case, we need to examine flow in three dimensions. Similarly, wellbores are not always vertical; in many oil fields, wells are frequently "inclined" and, in some cases, even horizontal. In such situations, we must, again, contend with flow in three dimensions. In geologic media that are not very permeable, it is not unusual to extend the surface area available for withdrawal by creating cracks known as hydraulic fractures. Such cracks are usually filled with sand and can be treated as a porous medium with properties distinct from those of the reservoir rock. Because of the prevailing stresses in the reservoir rock, such cracks are usually vertical, although horizontal cracks may develop at shallow depths. Hydraulic fractures or cracks are usually considered to be rectangular or circular sources.

The purpose of this section is to examine the details that are pertinent to the extraction of fluids for the conditions noted here, and then develop pressure distributions for a few cases that will permit us to discuss the essential characteristics of these solutions, so that efficient algorithms may be developed for computational purposes.

With the fundamental solutions derived in Chapter II, obtaining pressure distributions in porous media when fluid is extracted via lines or planes, in most cases, is a simple matter. If the strength of a continuous, point source located at M' is $\tilde{q}(M';t)/(\phi c)$, then the pressure distribution in terms of the Laplace transform of $\Delta p(M;t)$ is

$$\overline{\Delta p}(x_D, y_D, z_D) = \frac{\mu}{k\ell} \int_{\widetilde{S}} \bar{\tilde{q}}(x'_D, y'_D, z'_D)\,\overline{\gamma}\,(x_D - x'_D, y_D - y'_D, z_D - z'_D)\,dS', \quad (1)$$

where dS' denotes the element of a line or surface through which fluid is withdrawn. If we assume that the source-strength is uniform over time and space, then (1) becomes

$$\overline{\Delta p}(x_D, y_D, z_D) = \frac{\mu\tilde{q}}{k\ell s} \int_{\widetilde{S}} \overline{\gamma}\,(x_D - x'_D, y_D - y'_D, z_D - z'_D)\,dS'. \quad (2)$$

Solutions that satisfy the constraint in (2) are known as the *uniform-flux* solutions. In some cases it is possible to compute the pressure distribution on the source. In such cases we will find that the pressure will vary over the source. In many other cases, we are interested in imposing the constraint that the pressure distribution (on the source) be a function of time, that is, $\Delta p(\widetilde{M};t) = f(t)$. Such solutions will be referred to as *infinite-conductivity* solutions. We will discuss approximate procedures for obtaining infinite-conductivity solutions from the uniform-flux solutions. We shall now document solutions for a few systems.

3.1. Pressure distribution in slab reservoirs

In this section we explore the extraction of fluid from a variety of surfaces of interest to us. The fundamental solutions in Chapter 2 will be our starting point. Solutions given here assume that the top and the bottom of the reservoir are impermeable. The fundamental solution, $\overline{\gamma}$, for this system is given in 2.2(7). Solutions for the other boundary conditions readily follow and are given in Ozkan [1].

I. *Withdrawal via a rectangular plane source, perpendicular to the $z = 0$ plane with its center at (x_w, y_w, z_w) and parallel to the x-axis*

Let $2L_f$ be the length of the source and h_f be its height. If $\widetilde{q}$ is the flux, then the pressure distribution is obtained by substituting the right-hand side of 2.2(7) in 3(1) for $\overline{\gamma}$ and integrating with respect to z' from $z_w - h_f/2$ to $z_w + h_f/2$ and with respect to x' from $x_w - L_f$ to $x_w + L_f$. The appropriate expression is

$$\overline{\Delta p} = \frac{\mu\ell}{2\pi k z_{eD}} \int_{-h_f/(2\ell)}^{+h_f/(2\ell)} \int_{-L_f/\ell}^{+L_f/\ell} \overline{\widetilde{q}}(\widehat{x}_{wD}, \widehat{z}_{wD}) \Bigg\{ K_0\left[\sqrt{u}\sqrt{(x_D-\widehat{x}_{wD})^2+(y_D-y_{wD})^2}\right] + 2\sum_{n=1}^{\infty} K_0\left[\sqrt{u+\frac{n^2\pi^2}{z_{eD}^2}}\sqrt{(x_D-\widehat{x}_{wD})^2+(y_D-y_{wD})^2}\right]\cos n\pi\frac{z_D}{z_{eD}} \cos n\pi\frac{\widehat{z}_{wD}}{z_{eD}}\Bigg\} d\alpha d\beta. \tag{1}$$

Here, ℓ is the reference length, $\widehat{x}_{wD} = x_{wD} + \alpha\sqrt{k/k_x}$, $\widehat{z}_{wD} = z_{wD} + \beta\sqrt{k/k_z}$, and $u(s)$ is the Laplace transform variable with respect to normalized time, t_D. If q is the withdrawal rate from the porous medium, then $q = \int_{z_w-h_f/2}^{z_w+h_f/2}\int_{x_w-L_f}^{x_w+L_f} \widetilde{q}(\alpha,\beta)d\alpha d\beta$.

Several limiting forms of (1) can be derived. (Of course, they may also be obtained directly from the fundamental solution.) If the height of the source, h_f, is equal to the slab thickness $h \equiv |z_e|$, then $\widetilde{q}$ is uniform in z, and the pressure distribution is given by

$$\overline{\Delta p} = \frac{\mu h}{2\pi k z_{eD}} \int_{-L_f/\ell}^{+L_f/\ell} \overline{\widetilde{q}}(\widehat{x}_{wD}) K_0\left[\sqrt{u}\sqrt{(x_D-\widehat{x}_{wD})^2+(y_D-y_{wD})^2}\right] d\alpha. \tag{2}$$

If, however, fluid is extracted via a line source of length L_h, that is parallel to the

x-axis with its center at (x_w, y_w, z_w), then the pressure distribution is

$$\overline{\Delta p} = \frac{\mu}{2\pi k z_{eD}} \left\{ \int_{-L_h/(2\ell)}^{+L_h/(2\ell)} \overline{\tilde{q}}(\hat{x}_{wD}) K_0 \left[\sqrt{u} \sqrt{(x_D - \hat{x}_{wD})^2 + (y_D - y_{wD})^2} \right] d\alpha \right.$$
$$+ 2 \sum_{n=1}^{\infty} \cos n\pi \frac{z_D}{z_{eD}} \cos n\pi \frac{z_{wD}}{z_{eD}} \tag{3}$$
$$\left. \int_{-L_h/(2\ell)}^{+L_h/(2\ell)} \overline{\tilde{q}}(\hat{x}_{wD}) K_0 \left[\sqrt{u + \frac{n^2\pi^2}{z_{eD}^2}} \sqrt{(x_D - \hat{x}_{wD})^2 + (y_D - y_{wD})^2} \right] d\alpha \right\}.$$

Here, $\tilde{q}$ may be viewed as the flux.

It is convenient to express the solutions in terms of a normalized pressure given by

$$p_D(x_D, y_D, z_D; t_D) = \frac{2\pi k h}{q\mu} \left[p_i - p(x, y, z; t) \right], \tag{4}$$

where the pressure at $t = 0$ is p_i, h is the thickness ($h \equiv |z_e|$) and the withdrawal rate, q, is constant. In terms of (4), (1) may be written as (with $\ell = L_f$)

$$\overline{p}_D = \frac{1}{2h_{fD}} \int_{-h_{fD}\sqrt{k_z/k}/2}^{+h_{fD}\sqrt{k_z/k}/2} \int_{-1}^{+1} \overline{\tilde{q}}_{fD}(\hat{x}_{wD}, \hat{z}_{wD})$$
$$\left\{ K_0 \left[\sqrt{u} \sqrt{(x_D - \hat{x}_{wD})^2 + (y_D - y_{wD})^2} \right] \right. \tag{5}$$
$$+ 2 \sum_{n=1}^{\infty} K_0 \left[\sqrt{u + \frac{n^2\pi^2}{z_{eD}^2}} \sqrt{(x_D - \hat{x}_{wD})^2 + (y_D - y_{wD})^2} \right] \cos n\pi \frac{z_D}{z_{eD}}$$
$$\left. \cos n\pi \frac{\hat{z}_{wD}}{z_{eD}} \right\} d\alpha d\beta.$$

Here, $h_{fD} = h_f \sqrt{k/k_z}/L_f$ and $\tilde{q}_{fD} = (2L_f h_f \tilde{q})/q$, where $\tilde{q}$ may be viewed as the flux.

II. *Withdrawal from a hollow disc of radii r_1 and r_2 ($r_1 < r_2$) with its center at (x', y', z') and parallel to the plane $z = 0$*

We use the fundamental solution in 2.2(7) with the restriction that $k_x = k_y = k_r$. In this case, $k = \sqrt[3]{k_r^2 k_z}$.

Consider

$$K_0 \left[\sqrt{\beta} \sqrt{(x_D - x'_D)^2 + (y_D - y'_D)^2} \right].$$

Using the integral representation of $K_0(z)$,

$$K_0(z) = \frac{1}{2} \int_0^{\infty} \exp\left(-\xi - \frac{z^2}{4\xi} \right) \frac{d\xi}{\xi}; \ \mathcal{R}e(z^2) > 0, \tag{6}$$

and working in terms of polar coordinates, we have

$$
\begin{aligned}
&\int_0^{2\pi} K_0 \left\{ \sqrt{\beta} \left[r_D^2 + r_D'^2 - 2r_D r_D' \cos(\theta - \theta') \right]^{\frac{1}{2}} \right\} r_D' d\theta' \\
&\quad = \pi r_D' \int_0^{\infty} \exp(-\xi) \exp\left(-\frac{r_D^2 + r_D'^2}{4\xi} \beta \right) I_0 \left(\frac{r_D r_D'}{2\xi} \beta \right) \frac{d\xi}{\xi} \\
&\quad = \pi r_D' \int_0^{\infty} \exp(-\gamma\beta) \exp\left(-\frac{r_D^2 + r_D'^2}{4\gamma} \right) I_0 \left(\frac{r_D r_D'}{2\gamma} \right) \frac{d\gamma}{\gamma} \\
&\quad = \begin{cases} 2\pi r_D' I_0 \left(\sqrt{\beta} r_D' \right) K_0 \left(\sqrt{\beta} r_D \right); & r_D > r_D' \\ 2\pi r_D' I_0 \left(\sqrt{\beta} r_D \right) K_0 \left(\sqrt{\beta} r_D' \right); & r_D < r_D' \end{cases}.
\end{aligned}
\tag{7}
$$

Thus, if we consider a ring source in a slab reservoir and assume that $\bar{\tilde{q}}$ is independent of θ, then the pressure distribution is given by

$$
\begin{aligned}
\overline{\Delta p} = \frac{\mu \ell^2}{k z_e} &\left\{ \int_{r_{D1}}^{r_D} \bar{\tilde{q}} r_D' I_0 \left(\sqrt{u} r_D' \right) K_0 \left(\sqrt{u} r_D \right) dr_D' \right. \\
&+ \int_{r_D}^{r_{D2}} \bar{\tilde{q}} r_D' I_0 \left(\sqrt{u} r_D \right) K_0 \left(\sqrt{u} r_D' \right) dr_D' \\
&+ 2 \sum_{n=1}^{\infty} \cos n\pi \frac{z_D}{z_{eD}} \cos n\pi \frac{z_D'}{z_{eD}} \left[\int_{r_{D1}}^{r_D} \bar{\tilde{q}} r_D' I_0 \left(\sqrt{u + \alpha_n} r_D' \right) K_0 \left(\sqrt{u + \alpha_n} r_D \right) dr_D' \right. \\
&\left. \left. + \int_{r_D}^{r_{D2}} \bar{\tilde{q}} r_D' I_0 \left(\sqrt{u + \alpha_n} r_D \right) K_0 \left(\sqrt{u + \alpha_n} r_D' \right) dr_D' \right] \right\},
\end{aligned}
\tag{8}
$$

where $\tilde{q}$ is the flux and $\alpha_n = n^2\pi^2/z_{eD}^2$. If q is the withdrawal rate from the porous medium, then $q = 2\pi \int_{r_1}^{r_2} r' \tilde{q}(r') dr'$.

Let $\epsilon_n = \sqrt{u + n^2\pi^2/z_{eD}^2}$, and $a_D = \pi(r_{D2}^2 - r_{D1}^2)k/k_r$ be the normalized area of the source. We will assume that the flux distribution is uniform. The pressure distribution is then given by

$$
\begin{aligned}
\bar{p}_D(r_D, z_D, s) = \frac{2\pi}{a_D s} &\left\{ \int_{r_{D1}}^{r_D} r_D' I_0 \left(\sqrt{u} r_D' \right) K_0 \left(\sqrt{u} r_D \right) dr_D' \right. \\
&+ \int_{r_D}^{r_{D2}} r_D' I_0 \left(\sqrt{u} r_D \right) K_0 \left(\sqrt{u} r_D' \right) dr_D' \\
&+ 2 \sum_{n=1}^{\infty} \cos n\pi \frac{z_D}{z_{eD}} \cos n\pi \frac{z_D'}{z_{eD}} \left[\int_{r_{D1}}^{r_D} r_D' I_0 \left(\epsilon_n r_D' \right) K_0 \left(\epsilon_n r_D \right) dr_D' \right. \\
&\left. \left. + \int_{r_D}^{r_{D2}} r_D' I_0 \left(\epsilon_n r_D \right) K_0 \left(\epsilon_n r_D' \right) dr_D' \right] \right\}.
\end{aligned}
\tag{9}
$$

III. *Withdrawal from a disc of radius ρ_f with its center at (x_w, y_w, z_w) and perpendicular to the plane $z = 0$ and at an angle ν with respect to the $x - z$ plane (isotropic medium)*

Let ρ and ρ' be the distances of points $P(x, y, z)$ and $P'(x', y', z')$ from the point $O(x_w, y_w, z_w)$, respectively. If we now temporarily consider a coordinate system with the point O as the origin, and if (ρ, θ, ϕ) and (ρ', θ', ϕ') are the coordinates of the points P and P', respectively, then, for a porous medium that is infinite in extent, the fundamental solution in spherical coordinates is given by

$$\overline{\gamma} = \frac{1}{4\pi R_D} \exp(-\sqrt{u} R_D). \tag{10}$$

Here, $R_D^2 = |PP'|^2$, and by the law of cosines

$$R_D^2 = |PP'|^2 = \rho_D^2 + \rho_D'^2 - 2\rho_D \rho_D' \cos\varphi, \tag{11}$$

where φ is the angle POP' and

$$\cos\varphi = \cos\theta\cos\theta' + \sin\theta\sin\theta'\cos(\phi - \phi'). \tag{12}$$

Using the addition theorem of Bessel functions (see Watson [2] and Abramowitz and Stegun [3]), we have

$$\frac{\exp(-kR_D)}{R_D} = \begin{cases} \sum_{m=0}^{\infty} \frac{(2m+1)}{\sqrt{\rho_D \rho_D'}} F_m\left(k\rho_D', k\rho_D\right) P_m(\cos\varphi); & \rho_D < \rho_D' \\ \sum_{m=0}^{\infty} \frac{(2m+1)}{\sqrt{\rho_D \rho_D'}} F_m\left(k\rho_D, k\rho_D'\right) P_m(\cos\varphi); & \rho_D > \rho_D', \end{cases} \tag{13}$$

where $P_m(x)$ represents the Legendre polynomial and

$$F_m(a, b) = K_{m+1/2}(a) I_{m+1/2}(b). \tag{14}$$

Using (13), we can write (10) as

$$\overline{\gamma} = \frac{1}{4\pi} \sum_{m=0}^{\infty} \frac{(2m+1)}{\sqrt{\rho_D \rho_D'}} F\left(\sqrt{u}\rho_D', \sqrt{u}\rho_D\right) P_m(\cos\varphi), \tag{15}$$

where, for notational simplicity, we have defined

$$F(a, b) = \begin{cases} F_m(a, b) & b < a \\ F_m(b, a) & b > a. \end{cases} \tag{16}$$

If we now use the method of images and let $\overline{\gamma}_n$ denote the solution for the nth image in an infinite array of sources, then the fundamental solution for a slab reservoir with impermeable boundaries at $z_D = 0$ and $z_D = z_{eD}$ can be written as (see Appendix)

$$\overline{\gamma} = \sum_{n=-\infty}^{+\infty} \overline{\gamma}_n(\rho_{Dn}', \rho_{Dn}, \varphi_n). \tag{17}$$

Relating ρ'_{Dn}, ρ_{Dn}, and φ_n with ρ'_D, ρ_D, and φ, respectively, and then recasting (17), we obtain

$$\begin{aligned}\overline{\gamma} =& \frac{1}{4\pi\sqrt{\rho'_D}} \sum_{n=-\infty}^{+\infty} \sum_{m=0}^{+\infty} (2m+1) \left[\frac{F(\sqrt{u}\rho'_D, \sqrt{u}\rho_{Dn1})}{\sqrt{\rho_{Dn1}}} P_m(\cos\varphi_{n1}) \right. \\ &+ \left. \frac{F(\sqrt{u}\rho'_D, \sqrt{u}\rho_{Dn2})}{\sqrt{\rho_{Dn2}}} P_m(\cos\varphi_{n2}) \right],\end{aligned} \tag{18}$$

where

$$\rho^2_{Dn1} = (\rho_D \cos\theta + 2z_{wD} - 2nz_{eD})^2 + \rho^2_D \sin^2\theta, \tag{19}$$

$$\rho^2_{Dn2} = (\rho_D \cos\theta - 2nz_{eD})^2 + \rho^2_D \sin^2\theta, \tag{20}$$

$$\cos\varphi_k = \cos\theta_k \cos\theta'_k + \sin\theta_k \sin\theta'_k \cos(\phi - \phi'); \;\; k = n1 \text{ or } n2 \tag{21}$$

$$\theta_{n1} = \arccos \frac{\rho_D \cos\theta + 2z_{wD} - 2nz_{eD}}{\rho_{Dn1}}, \tag{22}$$

$$\theta_{n2} = \arccos \frac{\rho_D \cos\theta - 2nz_{eD}}{\rho_{Dn2}}, \tag{23}$$

$$\theta'_{n1} = \pi - \theta', \tag{24}$$

and

$$\theta'_{n2} = \theta'. \tag{25}$$

The pressure distribution owing to a disc source of radius $\rho_{fD} = \rho_f/\ell$, at an angle ν with respect to the $x - z$ plane and perpendicular to the plane $z = 0$ can now be written, by using (18), as follows (assuming uniform flux):

$$\overline{\Delta p} = \frac{\tilde{q}\mu\ell}{4\pi ks} \int_0^{\rho_{fD}} \sqrt{\rho'_D} \sum_{n=-\infty}^{+\infty} \sum_{m=0}^{+\infty} (2m+1)(G_{n1} + G_{n2}) d\rho'_D, \tag{26}$$

where

$$G_k = \frac{F(\sqrt{u}\rho'_D, \sqrt{u}\rho_{Dk})}{\sqrt{\rho_{Dk}}} \int_0^{\pi} [P_m(\cos\varphi_k)_{\phi'=\nu} + P_m(\cos\varphi_k)_{\phi'=\nu+\pi}] \, d\theta', \tag{27}$$

for $k = n1$ or $k = n2$. Note that by the addition theorem of associated Legendre functions (see Gradshteyn and Ryzhik [4], 8.814),

$$\begin{aligned}P_m(\cos\varphi) &= P_m[\cos\theta\cos\theta' + \sin\theta\sin\theta'\cos(\phi - \phi')] \\ &= P_m(\cos\theta)P_m(\cos\theta') \\ &+ 2\sum_{k=1}^{m} \frac{(m-k)!}{(m+k)!} \mathrm{P}^k_m(\cos\theta)\mathrm{P}^k_m(\cos\theta')\cos(\phi - \phi'); \;\; [0 \le \theta, \theta' \le \pi],\end{aligned} \tag{28}$$

where $\mathrm{P}_m^k(x)$ represents the associated Legendre function, we can write

$$P_m(\cos\varphi)_{\phi'=\nu} + P_m(\cos\varphi)_{\phi'=\nu+\pi} = 2P_m(\cos\theta)P_m(\cos\theta') + 2\sum_{k=1}^{m}\frac{(m-k)!}{(m+k)!}\mathrm{P}_m^k(\cos\theta)\mathrm{P}_m^k(\cos\theta')\left[\cos k(\phi-\nu)+\cos k(\phi-\nu-\pi)\right]. \tag{29}$$

Also noting that (see Gradshteyn and Ryzhik [5], 8.820)

$$P_\nu(\cos\theta) = \mathrm{P}_\nu(\cos\theta), \tag{30}$$

and using the result given in Gradshteyn and Ryzhik [6], 7.132.1, 7.166,

$$\begin{aligned}\int_0^\pi \mathrm{P}_m(\cos\theta')d\theta' &= \int_{-1}^{+1}\frac{\mathrm{P}_m(x)}{\sqrt{1-x^2}}dx^2 \\ &= \frac{\pi^2}{\left[\Gamma\left(1+\frac{m}{2}\right)\Gamma\left(\frac{1}{2}-\frac{m}{2}\right)\right]^2},\end{aligned} \tag{31}$$

we may write the normalized pressure as follows:

$$\begin{aligned}\bar{p}_D = \frac{z_{eD}}{\pi\rho_{fD}^2 s}\int_0^{\rho_{fD}}\sqrt{\rho'_D}\sum_{n=-\infty}^{+\infty}\sum_{m=0}^{+\infty}(2m+1)\Bigg\{&\left[\frac{\pi}{\Gamma\left(1+\frac{m}{2}\right)\Gamma\left(\frac{1}{2}-\frac{m}{2}\right)}\right]^2 \\ &\left[\frac{F(\sqrt{u}\rho'_D,\sqrt{u}\rho_{Dn1})}{\sqrt{\rho_{Dn1}}}\mathrm{P}_m(\cos\theta_{n1}) + \frac{F(\sqrt{u}\rho'_D,\sqrt{u}\rho_{Dn2})}{\sqrt{\rho_{Dn2}}}\mathrm{P}_m(\cos\theta_{n2})\right] \\ &+\sum_{k=1}^{m}\frac{(m-k)!}{(m+k)!}\left[\frac{F(\sqrt{u}\rho'_D,\sqrt{u}\rho_{Dn1})}{\sqrt{\rho_{Dn1}}}\mathrm{P}_m^k(\cos\theta_{n1})\right. \\ &\left.+\frac{F(\sqrt{u}\rho'_D,\sqrt{u}\rho_{Dn2})}{\sqrt{\rho_{Dn2}}}\mathrm{P}_m^k(\cos\theta_{n2})\right] \\ &\left[\cos k(\phi-\nu)+\cos k(\phi-\nu-\pi)\right]\int_0^\pi \mathrm{P}_m^k(\cos\theta')d\theta'\Bigg\}d\rho'_D,\end{aligned} \tag{32}$$

where we have used $q = \pi\rho_f^2\tilde{q}$.

IV. *Withdrawal from a line source of length h_f with its center at (x_w, y_w, z_w) and inclined at an angle ψ with respect to the vertical (in an isotropic medium)*

For illustrative purposes, we derive the solution in Cartesian and cylindrical coordinates.

A. Solution in Cartesian coordinates

If we consider a reference frame (x', y', z') that is obtained by rotating the x and z axes about the y axis in the clockwise direction by an angle ψ, then $x' = x\cos\psi - z\sin\psi$, $y' = y$, and $z' = x\sin\psi + z\cos\psi$.

For an infinite medium, the pressure distribution is given by

$$
\begin{aligned}
\overline{\Delta p} &= \frac{\mu}{4\pi k\ell}\int_{z'_w-h_f/2}^{z'_w+h_f/2} \bar{\bar{q}}(x'_{wD}, y'_{wD}, z''_{wD}) \frac{\exp\left[-\sqrt{u}\sqrt{r'^2_D+(z'_D-z''_{wD})^2}\right]}{\sqrt{r'^2_D+(z'_D-z''_{wD})^2}} dz''_w \\
&= \frac{\mu}{4\pi k}\int_{-h_f/(2\ell)}^{+h_f/(2\ell)} \bar{\bar{q}}(x'_{wD}, y'_{wD}, z'_{wD}-\xi) \frac{\exp\left[-\sqrt{u}\sqrt{r'^2_D+(z'_D-z'_{wD}-\xi)^2}\right]}{\sqrt{r'^2_D+(z'_D-z'_{wD}-\xi)^2}} d\xi \\
&= \frac{\mu}{4\pi k}\int_{-h_f/(2\ell)}^{+h_f/(2\ell)} \bar{\bar{q}}(\tilde{x}_{wD}, y_{wD}, \tilde{z}_{wD}) \frac{\exp\left[-\sqrt{u}\sqrt{\tilde{r}^2_D+(z_D-\tilde{z}_{wD})^2}\right]}{\sqrt{\tilde{r}^2_D+(z_D-\tilde{z}_{wD})^2}} d\xi,
\end{aligned}
\tag{33}
$$

where $r'^2_D = (x'_D - x'_{wD})^2 + (y'_D - y'_{wD})^2$, $\tilde{r}^2_D = (x_D - \tilde{x}_{wD})^2 + (y_D - y_{wD})^2$, $\tilde{x}_{wD} = x_{wD} + \xi \sin\psi$, and $\tilde{z}_{wD} = z_{wD} + \xi\cos\psi$. The solution for a slab reservoir can be obtained from (33) by using the method of images along the lines suggested in §2.2, and the pressure distribution is given by

$$
\begin{aligned}
\overline{\Delta p} = \frac{\mu}{2\pi k z_{eD}}\int_{-h_f/(2\ell)}^{+h_f/(2\ell)} \bar{\bar{q}}\Big[&K_0\left(\tilde{r}_D\sqrt{u}\right) \\
&+ 2\sum_{n=1}^{\infty} K_0\left(\tilde{r}_D\sqrt{u+\alpha_n}\right)\cos n\pi\frac{z_D}{z_{eD}}\cos n\pi\frac{\tilde{z}_{wD}}{z_{eD}}\Big] d\xi,
\end{aligned}
\tag{34}
$$

where $\alpha_n = n^2\pi^2/z^2_{eD}$.

The solution given in (18) is especially convenient if the line-source well is inclined at an angle ψ with respect to the vertical and tilted at an angle ν with respect to the $x - z$ plane. Using the fundamental solution given in (18) and assuming uniform flux, we obtain the following solution:

$$
\begin{aligned}
\overline{\Delta p} &= \frac{\tilde{q}\mu}{k\ell s}\int_0^{h_f/2} \left[\bar{\gamma}(\theta'=\psi, \phi'=\nu) + \bar{\gamma}(\theta'=\psi, \phi'=\nu+\pi)\right] d\rho' \\
&= \frac{\tilde{q}\mu}{4\pi k s}\sum_{n=-\infty}^{+\infty}\sum_{m=0}^{+\infty}(2m+1)\int_0^{h_f/(2\ell)} (G_{n1}+G_{n2})\frac{d\rho'_D}{\sqrt{\rho'_D}},
\end{aligned}
\tag{35}
$$

where for $k = n1$ or $k = n2$

$$
G_k = \frac{F\left(\sqrt{u}\rho'_D, \sqrt{u}\rho_{Dk}\right)}{\sqrt{\rho_{Dk}}}\left[P_m(\cos\varphi_k)_{\theta'=\psi,\phi'=\nu} + P_m(\cos\varphi_k)_{\theta'=\psi,\phi'=\nu+\pi}\right]. \tag{36}
$$

B. Solution in cylindrical coordinates

If the well is in the vertical $(x - z)$ plane (the case for $\nu = 0$ in Part A), then the solution can also be constructed in cylindrical coordinates. Let the cylindrical coordinates of the points (x, z, y) and (x', z', y') be given, respectively, by (r, θ, y) and (r', θ', y'). The fundamental solution for an infinite medium is given by

$$
\bar{\gamma} = \frac{1}{4\pi R_D}\exp(-\sqrt{u}R_D), \tag{37}
$$

where

$$
\begin{aligned}
R_D^2 &= (x_D - x'_D)^2 + (z_D - z'_D)^2 + (y_D - y'_D)^2 \\
&= r_D^2 + r_D'^2 - 2r_D r'_D \cos(\theta - \theta') + (y_D - y'_D)^2.
\end{aligned} \tag{38}
$$

Therefore, the fundamental solution for a slab reservoir with impermeable boundaries at $z_D = 0$ and $z_D = z_{eD}$ can be written, by using the method of images, as follows:

$$
\begin{aligned}
\overline{\gamma} =& \frac{1}{4\pi} \sum_{n=-\infty}^{+\infty} \left\{ \frac{\exp\left[-\sqrt{u}\sqrt{\widetilde{w}^2 + (\widetilde{z}_{D1} - 2nz_{eD})^2}\right]}{\sqrt{\widetilde{w}^2 + (\widetilde{z}_{D1} - 2nz_{eD})^2}} \right. \\
&+ \left. \frac{\exp\left[-\sqrt{u}\sqrt{\widetilde{w}^2 + (\widetilde{z}_{D2} - 2nz_{eD})^2}\right]}{\sqrt{\widetilde{w}^2 + (\widetilde{z}_{D2} - 2nz_{eD})^2}} \right\},
\end{aligned} \tag{39}
$$

where

$$
\widetilde{w}^2 = \widetilde{x}_D^2 + \widetilde{y}_D^2, \tag{40}
$$

$$
\widetilde{x}_D = x_D - x'_D = r_D \sin\theta - r'_D \sin\theta', \tag{41}
$$

$$
\widetilde{y}_D = y_D - y'_D = y_D - y_{wD}, \tag{42}
$$

$$
\widetilde{z}_{D1} = z_D - z'_D = r_D \cos\theta - r'_D \cos\theta' \tag{43}
$$

$$
\widetilde{z}_{D2} = z_D + z'_D = r_D \cos\theta - r'_D \cos(\theta' + \pi) + 2z_{wD}. \tag{44}
$$

Using the summation formula given in 2.2(6), we may recast (39) into the following form:

$$
\begin{aligned}
\overline{\gamma} =& \frac{1}{2\pi z_{eD}} \left[K_0\left(\sqrt{u}\widetilde{w}\right) + 2\sum_{n=1}^{\infty} K_0\left(\sqrt{u + \frac{n^2\pi^2}{z_{eD}^2}}\widetilde{w}\right) \right. \\
& \left. \cos n\pi \frac{z_{wD} + r_D \cos\theta}{z_{eD}} \cos n\pi \frac{z_{wD} + r'_D \cos\theta'}{z_{eD}} \right].
\end{aligned} \tag{45}
$$

The pressure distribution for an inclined line-source is now given by (assuming uniform flux)

$$
\begin{aligned}
\overline{\Delta p} =& \frac{\widetilde{q}\mu}{ks} \int_0^{h_f/(2\ell)} [\overline{\gamma}(\theta' = \psi) + \overline{\gamma}(\theta' = \psi + \pi)]\, dr'_D \\
=& \frac{\widetilde{q}\mu}{ks} \int_{-h_f/(2\ell)}^{+h_f/(2\ell)} \overline{\gamma}(\theta' = \psi) dr'_D \\
=& \frac{\widetilde{q}\mu}{2\pi k z_{eD} s} \int_{-h_f/(2\ell)}^{+h_f/(2\ell)} \left[K_0\left(\sqrt{u}\widetilde{w}\right) + 2\sum_{n=1}^{\infty} K_0\left(\sqrt{u + \frac{n^2\pi^2}{h_D^2}}\widetilde{w}\right) \right. \\
& \left. \cos n\pi \frac{z_{wD} + r_D \cos\theta}{z_{eD}} \cos n\pi \frac{z_{wD} + r'_D \cos\theta'}{z_{eD}} \right]_{\theta' = \psi} dr'_D.
\end{aligned} \tag{46}
$$

Note that

$$
\widetilde{w}^2 = \widetilde{r}_D^2 + r_D'^2 \sin^2\theta' - 2\widetilde{r}_D r'_D \sin\theta' \cos\xi, \tag{47}
$$

where $\widetilde{r}_D^2 = (x_D - x_{wD})^2 + (y_D - y_{wD})^2$, $z_{wD} + r_D \cos\theta = z_D$, and $\widetilde{r}_D \cos\xi = |x_D - x_{wD}|$; thus, the solution obtained in (46) is the same as that given in (34). It is also possible to use the addition theorem of Bessel functions given in 2.3(5) to express (46) in the following form:

$$\overline{\Delta p} = \frac{\widetilde{q}\mu}{2\pi k z_{eD} s} \int_{-h_f/(2\ell)}^{+h_f/(2\ell)} \sum_{k=-\infty}^{+\infty} \left[F_k(\sqrt{u}) + 2 \sum_{n=1}^{\infty} F_k\left(\sqrt{u + \frac{n^2\pi^2}{z_{eD}^2}}\right) \right.$$
$$\left. \cos n\pi \frac{z_{wD} + r_D \cos\theta}{z_{eD}} \cos n\pi \frac{z_{wD} + r'_D \cos\psi}{z_{eD}} \right] dr'_D, \tag{48}$$

where

$$F_k(a) = \begin{cases} I_k(a\widetilde{r}_D) K_k(ar'_D \sin\psi) \cos\xi; & \widetilde{r}_D < r'_D \sin\psi \\ I_k(ar'_D \sin\psi) K_k(a\widetilde{r}_D) \cos\xi; & \widetilde{r}_D > r'_D \sin\psi. \end{cases} \tag{49}$$

V. *Fluid withdrawal from a rectangular region with properties distinct from the porous medium*

Here, we examine fluid extraction via a vertical fracture as discussed in the Introduction to this chapter. Let us assume that the vertical fracture can be represented by a rectangle in the region $-L_f < x < L_f$, $-w_f/2 < y < w_f/2$, $(L_f >> w_f)$ with the center of the rectangle at the $(0, 0, z_e/2)$. The height of the fracture is equal to the thickness of the porous medium. We assume that fluid is withdrawn from the fracture at a rate q over a strip (line) of width w_f that is centered at the origin. Fluid enters the fracture only along the longer sides of the rectangle; however, the flux distribution is unspecified; that is, the flux distribution at $|y| = w_f/2$ is unknown and must be determined.

Because $L_f >> w_f$ we will assume that pressure gradients in the y-direction are negligibly small and thus consider flow only in the x-direction within the fracture. Also, flow within the fracture will be considered to be steady because the volume of the fracture is negligibly small compared with the porous medium that surrounds it. Because of symmetry, we restrict our attention to the first quadrant. We follow the ideas of Cinco and Meng [7].

If $p_f(x;t)$ denotes the pressure distribution in the fracture and p_{fD} is given by (4), then we can readily show that pressure distribution is given by

$$\frac{d^2\overline{p}_{fD}}{dx_D^2} - \frac{\pi}{C_{fD}} \overline{\widetilde{q}}_D(x_D) = 0. \tag{50}$$

Here, $C_{fD} = k_f w/(kL_f)$, $\widetilde{q}_D = 2L_f\widetilde{q}/q$ and the reference length is chosen to be L_f. Integration of (50), using the appropriate boundary conditions (outer boundary is sealed and fluid is withdrawn at a constant rate at $x = 0$; that is, $\partial p_{fD}/\partial x_D = -\pi/C_{fD}$ at $x_D = 0$), yields the pressure distribution within the fracture, given by the following expression:

$$\overline{p}_{wD} - \overline{p}_{fD} = \frac{\pi}{C_{fD} s} \left[x_D - s \int_0^{x_D} \int_0^{x'} \overline{\widetilde{q}}_D(x'') dx'' dx' \right]. \tag{51}$$

Here, p_{wD} is the normalized pressure at $x_D = 0$.

The pressure distribution in the porous medium (assuming the porous medium is an isotropic system) is given by ($\widetilde{q}_D \equiv \widetilde{q}$)

$$\overline{p}_D(x_D, y_D, s) = \frac{1}{2}\int_{-1}^{+1} \overline{\widetilde{q}}(\alpha, s) K_0 \left\{ \left[(x_D - \alpha)^2 + y_D^2 \right]^{\frac{1}{2}} \sqrt{u} \right\} d\alpha. \tag{52}$$

Using (52) for $\overline{p}_{fD}$ in (51), we have

$$\begin{aligned} \overline{p}_{wD} - \frac{1}{2}\int_0^1 \overline{\widetilde{q}}(x', s) \left\{ K_0\left(|x_D - x'|\sqrt{u}\right) + K_0\left[(x_D + x')\sqrt{u}\right] \right\} dx' \\ + \frac{\pi}{C_{fD}} \int_0^{x_D}\int_0^{x'} \overline{\widetilde{q}}(x'', s)\, dx'' dx' = \frac{\pi x_D}{C_{fD} s}. \end{aligned} \tag{53}$$

We shall now solve (53) numerically. Consider the partition $0 = x_{D1}, \ldots, x_{Di}, x_{Di+1}, \ldots, x_{Dn+1} = 1$. If we assume that the fracture may be divided into n segments, then we can write

$$\begin{aligned} &\int_0^1 \overline{\widetilde{q}}(x', s) \left\{ K_0\left(|x_D - x'|\sqrt{u}\right) + K_0\left[(x_D + x')\sqrt{u}\right] \right\} dx' \\ &= \sum_{i=1}^{n} \overline{\widetilde{q}}_i(s) \int_{x_{Di}}^{x_{Di+1}} \left\{ K_0\left(|x_D - x'|\sqrt{u}\right) + K_0\left[(x_D + x')\sqrt{u}\right] \right\} dx'. \end{aligned} \tag{54}$$

The second integral in (53) may be written as

$$\begin{aligned} \int_0^{x_{Dj}}\int_0^{x'} \overline{\widetilde{q}}(x'', s)\, dx'' dx' = \sum_{i=1}^{j-1} \overline{\widetilde{q}}_i(s) \left[\frac{\Delta x_D^2}{2} + \Delta x_D (x_{Dj} - i\Delta x_D) \right] \\ + \frac{\Delta x_D^2}{8} \overline{\widetilde{q}}_j(s), \end{aligned} \tag{55}$$

where x_{Dj} is the midpoint of the jth segment and Δx_D is the width of each segment. In addition to the above expressions, by virtue of steady flow, we require

$$\Delta x_D \sum_{i=1}^{n} \overline{\widetilde{q}}_i(s) = \frac{1}{s}. \tag{56}$$

(53)–(56) constitute the system of equations that needs to be solved to determine the unknowns, $\widetilde{q}_i(t_D)$ and p_{wD}. The principal advantage of this scheme is that the system can be solved for any time t, and such solutions are independent of the solutions for previous times.

VI. *Infinite-conductivity fracture*

The development in V permits us to outline the algorithm to compute pressure distributions for an infinite-conductivity fracture. The ideas given here were first used by Muskat [8] to study steady-flow problems. Consider (52),

$$\bar{p}_D = \frac{1}{2}\int_{-1}^{+1} \bar{\tilde{q}}(\alpha, s) K_0 \left\{ \left[(x_D - \alpha)^2 + y_D^2 \right]^{\frac{1}{2}} \sqrt{u} \right\} d\alpha.$$

We may now use (54) to replace the integral in (52). For an infinite-conductivity wellbore, we require that

$$\bar{p}_D(x_{Dj}) = \bar{p}_D(x_{Dj+1}), j = 1, \dots n-1. \tag{57}$$

With the additional constraint given in (56), we may solve the resulting system of equations to obtain the flux distribution and the wellbore pressure-response. We defer further discussion of this matter until Chapter IV.

3.2. Pressure distribution in cylindrical porous media

We consider two examples for illustrative purposes: a vertically-fractured well in a closed, circular cylinder, and a horizontal, line source in a composite region.

I. *A vertically-fractured well*

We assume that the porous medium is isotropic and the boundaries at $z = 0$, $z = z_e$, and $r = r_e$ are impermeable. We consider a vertically-fractured well of height $h \equiv |z_e|$ and length $2L_f$. The center of the fracture is at $(0, 0, z_e/2)$ and the fracture tips extend from $(L_f, \alpha + \pi)$ to (L_f, α). The flux distribution is assumed to be uniform.

The fundamental solution for this system is given in 2.3(10). If we use 2.3(10) in 3(1) for $\bar{\gamma}$ and integrate with respect to z' from $z_w - h/2$ to $z_w + h/2$, we obtain the pressure distribution owing to a vertical, line-source well located at the point (r'_D, θ') (see also Carslaw and Jaeger [9])

$$\begin{aligned}\overline{\Delta p} &= \frac{\tilde{q}\mu h}{2\pi k \ell z_{eD} s}\Bigg[K_0\left(\sqrt{u}R_D\right) \\ &\quad - \sum_{k=-\infty}^{+\infty} I_k\left(\sqrt{u}r_D\right) \frac{I_k\left(\sqrt{u}r'_D\right) K'_k\left(\sqrt{u}r_{eD}\right)}{I'_k\left(\sqrt{u}r_{eD}\right)} \cos k(\theta - \theta')\Bigg] \\ &= \frac{\tilde{q}\mu h}{2\pi k \ell z_{eD} s} \sum_{k=-\infty}^{+\infty} F_k \cos k(\theta - \theta').\end{aligned} \tag{1}$$

Here, $R_D^2 = r_D^2 + r_D'^2 - 2r_D r'_D \cos(\theta - \theta')$, and

$$F_k = \begin{cases} F_k(r_D, r'_D) & \text{for } r_D < r'_D \\ F_k(r'_D, r_D) & \text{for } r_D > r'_D \end{cases}, \tag{2}$$

where

$$F_k(a,b) = I_k\left(a\sqrt{u}\right)\left[\frac{K_k\left(b\sqrt{u}\right)I_k'\left(r_{eD}\sqrt{u}\right) - I_k\left(b\sqrt{u}\right)K_k'\left(r_{eD}\sqrt{u}\right)}{I_k'\left(r_{eD}\sqrt{u}\right)}\right]. \tag{3}$$

The second equality in (1) follows from the addition theorem for $K_0(\sqrt{u}R_D)$ given in 2.3(5).

To obtain the pressure distribution caused by a vertical fracture extending between the points $(L_f, \alpha + \pi)$ and (L_f, α), we integrate (1) with respect to r' as follows:

$$\begin{aligned}\overline{\Delta p} &= \frac{\tilde{q}\mu h}{2\pi k z_{eD} s}\sum_{k=-\infty}^{+\infty}\left[\cos k(\theta-\alpha-\pi)\int_0^{L_{fD}} F_k dr'_D + \cos k(\theta-\alpha)\int_0^{L_{fD}} F_k dr'_D\right] \\ &= \frac{\tilde{q}\mu h}{\pi k z_{eD} s}\sum_{k=-\infty}^{+\infty}\cos k\left(\theta-\alpha-\frac{\pi}{2}\right)\cos k\frac{\pi}{2}\int_0^{L_{fD}} F_k dr'_D.\end{aligned} \tag{4}$$

Here, $L_{fD} = L_f/\ell$. In terms of normalized pressure, (4) becomes (with $\ell = L_f$)

$$\overline{p}_D = \frac{1}{s}\sum_{k=-\infty}^{+\infty}\cos k\left(\theta-\alpha-\frac{\pi}{2}\right)\cos k\frac{\pi}{2}\int_0^1 F_k dr'_D, \tag{5}$$

where

$$\int_0^1 F_k dr'_D = \begin{cases}\int_0^1 F_k(r'_D, r_D)dr'_D & \text{for } r_D \geq 1 \\ \int_0^{r_D} F_k(r'_D, r_D)dr'_D + \int_{r_D}^1 F_k(r_D, r'_D)dr'_D & \text{for } r_D \leq 1\end{cases}. \tag{6}$$

(5) can also be written as follows:

$$\begin{aligned}\overline{p}_D = \overline{p}_{Di} - \frac{1}{s}\sum_{k=-\infty}^{+\infty}\cos k\left(\theta-\alpha-\frac{\pi}{2}\right)\cos k\frac{\pi}{2} \\ \frac{I_k\left(r_D\sqrt{u}\right)K_k'\left(r_{eD}\sqrt{u}\right)}{I_k'\left(r_{eD}\sqrt{u}\right)}\int_0^1 I_k\left(r'_D\sqrt{u}\right)dr'_D.\end{aligned} \tag{7}$$

Here, $\overline{p}_{Di}$ is the normalized pressure for a vertically-fractured well in a slab reservoir (see §3.1 I) and is given by

$$\begin{aligned}\overline{p}_{Di} &= \frac{1}{2s}\int_0^1\left\{K_0\left[\sqrt{u}\sqrt{r_D^2 + r_D'^2 - 2r_D r'_D\cos(\theta-\alpha)}\right]\right. \\ &\left. + K_0\left[\sqrt{u}\sqrt{r_D^2 + r_D'^2 - 2r_D r'_D\cos(\theta-\alpha-\pi)}\right]\right\}dr'_D \\ &= \frac{1}{2s}\sum_{k=-\infty}^{+\infty}\left[\cos k(\theta-\alpha-\pi) + \cos k(\theta-\alpha)\right]\int_0^1 F_{1k} dr'_D,\end{aligned} \tag{8}$$

where $\int_0^1 F_{1k} dr'_D$ is given by

$$\begin{cases} \int_0^1 I_k\left(r'_D\sqrt{u}\right) dr'_D K_k\left(r_D\sqrt{u}\right) & \text{for } r_D \geq 1 \\ \int_0^{r_D} I_k\left(r'_D\sqrt{u}\right) dr'_D K_k\left(r_D\sqrt{u}\right) + I_k\left(r_D\sqrt{u}\right) \int_{r_D}^1 K_k\left(r'_D\sqrt{u}\right) dr'_D & \text{for } r_D \leq 1. \end{cases}$$

II. *A horizontal, line source in a composite region*

From the fundamental solution given in 2.3(34) and 2.3(35), the pressure distribution caused by a horizontal, line-source well of length L_h extending between the points $(L_h/2, \alpha+\pi, z_w)$ and $(L_h/2, \alpha, z_w)$ and with its center at $r = 0$ and $z = z_w$, is given by (assuming uniform flux)

$$\overline{p}_D(r_D < a_D) = \frac{1}{s}\sqrt{\frac{k_{z1}}{k_1}} \int_0^1 \left[G\left(\theta' = \alpha + \pi, r'_D = \xi\sqrt{k_1/k_{r1}}\right) + G\left(\theta' = \alpha, r'_D = \xi\sqrt{k_1/k_{r1}}\right)\right] d\xi, \tag{9}$$

and

$$\overline{p}_D(r_D > a_D) = \frac{\lambda_{rD}}{a_D s}\sqrt{\frac{k_{z1}}{k_1}} \int_0^1 \left[H\left(\theta' = \alpha + \pi, r'_D = \xi\sqrt{k_1/k_{r1}}\right) + H\left(\theta' = \alpha, r'_D = \xi\sqrt{k_1/k_{r1}}\right) d\xi, \right. \tag{10}$$

where

$$G = \sum_{k=-\infty}^{+\infty} S_{k0} + 2\sum_{n=1}^{\infty} \cos n\pi \frac{z_D}{z_{eD}} \cos n\pi \frac{z_{wD}}{z_{eD}} \sum_{k=-\infty}^{+\infty} S_{kn}, \tag{11}$$

and

$$H = \sum_{k=-\infty}^{+\infty} R_{k0} + 2\sum_{n=1}^{\infty} \cos n\pi \frac{\widetilde{z}_D}{\widetilde{z}_{eD}} \cos n\pi \frac{\widetilde{z}_{wD}}{\widetilde{z}_{eD}} \sum_{k=-\infty}^{+\infty} R_{kn}. \tag{12}$$

In (11) and (12), S_{kn} and R_{kn} are defined by 2.3(36) and 2.3(37), respectively.

3.3. Pressure distribution in rectangular parallelepipeds

The ideas developed in §3.1 can be readily extended to porous media bounded in the x and y directions. To demonstrate the procedure we now consider example applications. For simplicity, we assume that the flux distribution is *uniform*. Also, here, $h \equiv |z_e|$.

I. *All sides of the porous solid are impermeable*

The fundamental solution for this system is given in 2.4(13). We consider two examples.

A. A vertical fracture of length $2L_f$ with its center at $(x_w, y_w, z_e/2)$.

Let h be the height of the fracture. The solution to this problem is obtained by substituting in 3(1) the expression for $\overline{\gamma}$ given by 2.4(13), and integrating the right-hand side of 3(1) with respect to z' over the interval $z_w - h/2$ to $z_w + h/2$, and with respect to x' over the interval $x_w - L_f$ to $x_w + L_f$. The expression for the pressure distribution is

$$\overline{\Delta p} = \frac{\tilde{q}\mu h L_f}{k\ell x_{eD} z_{eD} s}\left[\frac{ch\sqrt{u}(y_{eD} - |\tilde{y}_{D1}|) + ch\sqrt{u}(y_{eD} - \tilde{y}_{D2})}{\sqrt{u}\ sh\sqrt{u}\ y_{eD}}\right.$$
$$+ \frac{2x_e}{\pi L_f}\sum_{k=1}^{\infty}\frac{1}{k}\sin k\pi\frac{L_f}{x_e}\cos k\pi\frac{x_w}{x_e}\cos k\pi\frac{x}{x_e} \tag{1}$$
$$\left.\frac{ch\sqrt{u+\frac{k^2\pi^2}{x_{eD}^2}}(y_{eD} - |\tilde{y}_{D1}|) + ch\sqrt{u+\frac{k^2\pi^2}{x_{eD}^2}}(y_{eD} - \tilde{y}_{D2})}{\sqrt{u+\frac{k^2\pi^2}{x_{eD}^2}}\ sh\sqrt{u+\frac{k^2\pi^2}{x_{eD}^2}}y_{eD}}\right].$$

If we choose the characteristic length ℓ to be L_f and assume $k = k_x = k_y = k_z$, then, using the definition of $\overline{p}_D$ given by 3.1(4) (with $q = 2h\tilde{q}L_f$), we can write (1) as

$$\overline{p}_D(x_D, y_D) = \frac{\pi}{x_{eD}s}\left[\frac{ch\sqrt{u}(y_{eD} - |\tilde{y}_{D1}|) + ch\sqrt{u}(y_{eD} - \tilde{y}_{D2})}{\sqrt{u}sh\sqrt{u}y_{eD}}\right.$$
$$+ \frac{2x_{eD}}{\pi}\sum_{k=1}^{\infty}\frac{1}{k}\sin k\pi\frac{1}{x_{eD}}\cos k\pi\frac{x_{wD}}{x_{eD}}\cos k\pi\frac{x_D}{x_{eD}} \tag{2}$$
$$\left.\frac{ch\sqrt{u+\frac{k^2\pi^2}{x_{eD}^2}}(y_{eD} - |\tilde{y}_{D1}|) + ch\sqrt{u+\frac{k^2\pi^2}{x_{eD}^2}}(y_{eD} - \tilde{y}_{D2})}{\sqrt{u+\frac{k^2\pi^2}{x_{eD}^2}}sh\sqrt{u+\frac{k^2\pi^2}{x_{eD}^2}}y_{eD}}\right].$$

B. A line source of length L_h parallel to the plane $z = 0$ with its center at (x_w, y_w, z_w).

This solution is obtained by substituting in 3(1) the expression for $\overline{\gamma}$ in 2.4(13) and integrating the right-hand side of 3(1) with respect to x' from $x_w - L_h/2$ to $x_w + L_h/2$. Assuming the porous medium to be an isotropic system and using $\ell = L_h/2$, the normalized pressure is given by

$$\overline{p}_D(x_D, y_D, z_D) = \overline{p}_{fD}(x_D, y_D) + \overline{F}_1(x_D, y_D, z_D), \tag{3}$$

where $\overline{p}_D$ is defined by 3.1(4) (with $q = \tilde{q}L_h$), $\overline{p}_{fD}$ is the fracture solution given by the right-hand side of (2), and $\overline{F}_1$ is defined by

$$\overline{F}_1 = \frac{2\pi}{x_{eD}s}\sum_{n=1}^{\infty}\cos n\pi z_D\cos n\pi z_{wD}$$
$$\frac{ch\sqrt{u+n^2\pi^2L_D^2}(y_{eD} - |\tilde{y}_{D1}|) + ch\sqrt{u+n^2\pi^2L_D^2}(y_{eD} - \tilde{y}_{D2})}{\sqrt{u+n^2\pi^2L_D^2}\ sh\sqrt{u+n^2\pi^2L_D^2}y_{eD}}$$
$$+ \frac{4}{s}\sum_{n=1}^{\infty}\cos n\pi z_D\cos n\pi z_{wD}\sum_{k=1}^{\infty}\frac{1}{k}\sin k\pi\frac{1}{x_{eD}}\cos k\pi\frac{x_{wD}}{x_{eD}}\cos k\pi\frac{x_D}{x_{eD}} \tag{4}$$
$$\frac{ch\sqrt{u+\frac{k^2\pi^2}{x_{eD}^2}+n^2\pi^2L_D^2}(y_{eD} - |\tilde{y}_{D1}|) + ch\sqrt{u+\frac{k^2\pi^2}{x_{eD}^2}+n^2\pi^2L_D^2}(y_{eD} - \tilde{y}_{D2})}{\sqrt{u+\frac{k^2\pi^2}{x_{eD}^2}+n^2\pi^2L_D^2}sh\sqrt{u+\frac{k^2\pi^2}{x_{eD}^2}+n^2\pi^2L_D^2}y_{eD}}.$$

In (4), z_D and L_D are defined, respectively, by

$$z_D = z/z_e \tag{5}$$

and

$$L_D = \frac{L_h}{2z_e}. \tag{6}$$

II. *One boundary is at a constant (initial) pressure*

We consider an *inclined* well in an isotropic reservoir and assume that the constant-pressure boundary is at $x_D = x_{eD}$. The well length is assumed to be h_f and let $h_{fD} = h_f/\ell$. The center of the well is at (x_w, y_w, z_w) and it is inclined at an angle ψ to the vertical; see §3.1 IV. Following the ideas in §2.4 II, the pressure distribution is given by

$$\begin{aligned}\overline{\Delta p} = \frac{\widetilde{q}\mu}{4\pi k s} \int_{-h_{fD}/2}^{+h_{fD}/2} \sum_{k=-\infty}^{+\infty} \sum_{m=-\infty}^{+\infty} \sum_{n=-\infty}^{+\infty} (-1)^k (\widetilde{S}_{1,1,1} + \widetilde{S}_{2,1,1} + \widetilde{S}_{1,2,1} \\ + \widetilde{S}_{2,2,1} + \widetilde{S}_{1,1,2} + \widetilde{S}_{2,1,2} + \widetilde{S}_{1,2,2} + \widetilde{S}_{2,2,2}) d\xi,\end{aligned} \tag{7}$$

where

$$\widetilde{S}_{i,j,\ell} = \frac{\exp\left[-\sqrt{u}\sqrt{(\widehat{x}_{Di} - 2kx_{eD})^2 + (\widehat{y}_{Dj} - 2my_{eD})^2 + (\widehat{z}_{D\ell} - 2nz_{eD})^2}\right]}{\sqrt{(\widehat{x}_{Di} - 2kx_{eD})^2 + (\widehat{y}_{Dj} - 2my_{eD})^2 + (\widehat{z}_{D\ell} - 2nz_{eD})^2}}, \tag{8}$$

$$\widehat{x}_{Di} = x_D + (-1)^i \widetilde{x}_{wD} \tag{9}$$

$$\widehat{y}_{Dj} = y_D + (-1)^j y_{wD} \tag{10}$$

$$\widehat{z}_{D\ell} = z_D + (-1)^\ell \widetilde{z}_{wD}, \tag{11}$$

for $i, j, \ell = 1$ or 2. $\widetilde{x}_{wD}$ and $\widetilde{z}_{wD}$ are given, respectively, by

$$\widetilde{x}_{wD} = x_{wD} + \xi \sin\psi \tag{12}$$

and

$$\widetilde{z}_{wD} = z_{wD} + \xi \cos\psi. \tag{13}$$

In passing, we should note that if we replace $\widetilde{x}_{wD}$ by x_{wD} and $\widetilde{z}_{wD}$ by $z_{wD} + \xi$, then we will obtain the pressure distribution owing to a vertical well of length h_f. The converse is also true; that is, to obtain the pressure distribution for an inclined well, we can write the pressure distribution for a vertical well of length h_f and then replace x_{wD} and z_{wD} by $\widetilde{x}_{wD}$ and $\widetilde{z}_{wD}$, respectively.

Thus, as in §2.4 II, the expression for the pressure distribution is given by

$$\overline{p}_D(x_D, y_D, z_D) = \frac{2\pi}{x_{eD} h_{fD} s} \int_{-h_{fD}/2}^{+h_{fD}/2} \sum_{k=1}^{\infty} \cos(2k-1)\frac{\pi}{2}\frac{x_D}{x_{eD}} \cos(2k-1)\frac{\pi}{2}\frac{\widetilde{x}_{wD}}{x_{eD}}$$
$$\left[\frac{ch\sqrt{u + \frac{(2k-1)^2\pi^2}{4x_{eD}^2}}(y_{eD} - |\widetilde{y}_{D1}|) + ch\sqrt{u + \frac{(2k-1)^2\pi^2}{4x_{eD}^2}}(y_{eD} - \widetilde{y}_{D2})}{\sqrt{u + \frac{(2k-1)^2\pi^2}{4x_{eD}^2}}\, sh\sqrt{u + \frac{(2k-1)^2\pi^2}{4x_{eD}^2}}\, y_{eD}} \right.$$
$$+ 2\sum_{n=1}^{\infty} \cos n\pi \frac{z_D}{z_{eD}} \cos n\pi \frac{\widetilde{z}_{wD}}{z_{eD}}$$
$$\left. \frac{ch\sqrt{u + \frac{(2k-1)^2\pi^2}{4x_{eD}^2} + \frac{n^2\pi^2}{z_{eD}^2}}(y_{eD} - |\widetilde{y}_{D1}|) + ch\sqrt{u + \frac{(2k-1)^2\pi^2}{4x_{eD}^2} + \frac{n^2\pi^2}{z_{eD}^2}}(y_{eD} - \widetilde{y}_{D2})}{\sqrt{u + \frac{(2k-1)^2\pi^2}{4x_{eD}^2} + \frac{n^2\pi^2}{z_{eD}^2}}\, sh\sqrt{u + \frac{(2k-1)^2\pi^2}{4x_{eD}^2} + \frac{n^2\pi^2}{z_{eD}^2}}\, y_{eD}} \right] d\xi. \tag{14}$$

III. *Two boundaries are at a constant (initial) pressure*

We consider a vertical well of length h with its center at $(x_w, y_w, z_e/2)$ in an isotropic reservoir and assume that the constant-pressure boundaries are at $x_D = x_{eD}$ and $y_D = y_{eD}$. Using the appropriate fundamental solution, 2.4(28), in 3(1) and suitably integrating, we find that the pressure distribution is given by

$$\overline{\Delta p} = \frac{\widetilde{q}\mu}{k x_{eD} s} \sum_{k=1}^{\infty} \cos(2k-1)\frac{\pi}{2}\frac{x_D}{x_{eD}} \cos(2k-1)\frac{\pi}{2}\frac{x_{wD}}{x_{eD}}$$
$$\frac{sh\sqrt{u + \frac{(2k-1)^2\pi^2}{4x_{eD}^2}}(y_{eD} - |\widetilde{y}_{D1}|) + sh\sqrt{u + \frac{(2k-1)^2\pi^2}{4x_{eD}^2}}(y_{eD} - \widetilde{y}_{D2})}{\sqrt{u + \frac{(2k-1)^2\pi^2}{4x_{eD}^2}}\, ch\sqrt{u + \frac{(2k-1)^2\pi^2}{4x_{eD}^2}}\, y_{eD}}. \tag{15}$$

In terms of the normalized variables, (15) becomes

$$\overline{p}_D(x_D, y_D) = \frac{2\pi}{x_{eD} s} \sum_{k=1}^{\infty} \cos(2k-1)\frac{\pi}{2}\frac{x_D}{x_{eD}} \cos(2k-1)\frac{\pi}{2}\frac{x_{wD}}{x_{eD}}$$
$$\frac{sh\sqrt{u + \frac{(2k-1)^2\pi^2}{4x_{eD}^2}}(y_{eD} - |\widetilde{y}_{D1}|) + sh\sqrt{u + \frac{(2k-1)^2\pi^2}{4x_{eD}^2}}(y_{eD} - \widetilde{y}_{D2})}{\sqrt{u + \frac{(2k-1)^2\pi^2}{4x_{eD}^2}}\, ch\sqrt{u + \frac{(2k-1)^2\pi^2}{4x_{eD}^2}}\, y_{eD}}. \tag{16}$$

IV. *Four boundaries of the porous medium are at a constant (initial) pressure*

We will assume that the boundaries $z_D = 0$ and $z_D = z_{eD}$ are impermeable and all other boundaries are at a constant pressure. The fundamental solution for this system is given in 2.4(42). We assume that fluid is produced via a vertical fracture of length $2L_f$ and height h in a reservoir that is isotropic.

This solution can be obtained as in I. Integrating 3(1) with respect to z' from $z_w - h/2$ to $z_w + h/2$ after substituting the expression for $\overline{\gamma}$ given in 2.4(42)yields

$$\overline{\Delta p} = \frac{\widetilde{q}\mu}{kx_{eD}s} \sum_{k=1}^{\infty} \sin k\pi \frac{x_D}{x_{eD}} \sin k\pi \frac{x'_D}{x_{eD}}$$
$$\frac{ch\sqrt{u + \frac{k^2\pi^2}{x_{eD}^2}}(y_{eD} - |\widetilde{y}_{D1}|) - ch\sqrt{u + \frac{k^2\pi^2}{x_{eD}^2}}(y_{eD} - \widetilde{y}_{D2})}{\sqrt{u + \frac{k^2\pi^2}{x_{eD}^2}} sh\sqrt{u + \frac{k^2\pi^2}{x_{eD}^2}} y_{eD}}. \tag{17}$$

Further integration of (17) with respect to x' from $x_w - L_f$ to $x_w + L_f$ yields

$$\overline{\Delta p} = \frac{2L_f\widetilde{q}\mu}{\pi ks} \sum_{k=1}^{\infty} \frac{1}{k} \sin k\pi \frac{x_D}{x_{eD}}$$
$$\sin k\pi \frac{x_{wD}}{x_{eD}} \sin k\pi \frac{1}{x_{eD}} \frac{ch\sqrt{u + \frac{k^2\pi^2}{x_{eD}^2}}(y_{eD} - |\widetilde{y}_{D1}|) - ch\sqrt{u + \frac{k^2\pi^2}{x_{eD}^2}}(y_{eD} - \widetilde{y}_{D2})}{\sqrt{u + \frac{k^2\pi^2}{x_{eD}^2}} sh\sqrt{u + \frac{k^2\pi^2}{x_{eD}^2}} y_{eD}}. \tag{18}$$

In terms of $\overline{p}_D$, we have

$$\overline{p}_D(x_D, y_D) = \frac{2}{s} \sum_{k=1}^{\infty} \frac{1}{k} \sin k\pi \frac{x_D}{x_{eD}} \sin k\pi \frac{x_{wD}}{x_{eD}} \sin k\pi \frac{1}{x_{eD}}$$
$$\frac{ch\sqrt{u + \frac{k^2\pi^2}{x_{eD}^2}}(y_{eD} - |\widetilde{y}_{D1}|) - ch\sqrt{u + \frac{k^2\pi^2}{x_{eD}^2}}(y_{eD} - \widetilde{y}_{D2})}{\sqrt{u + \frac{k^2\pi^2}{x_{eD}^2}} sh\sqrt{u + \frac{k^2\pi^2}{x_{eD}^2}} y_{eD}}. \tag{19}$$

References

1. Ozkan. E.: *Performance of Horizontal Wells*, Ph.D. Dissertation, University of Tulsa (1988), 16–21, 247–249.
2. Watson, G. N.: *A Treatise on the Theory of Bessel Functions*, Cambridge University Press, London (1948), 366.
3. Abramowitz, M. and Stegun, I. A.: *Handbook of Mathematical Functions*, Dover Publications, Inc., New York (1972), 445.
4. Gradshteyn, I. S. and Ryzhik, I. M.: *Table of Integrals, Series, and Products*, Academic Press, Inc., Orlando (1980), 1015.
5. Gradshteyn, I. S. and Ryzhik, I. M.: *op. cit.*, 1016, 1017.
6. Gradshteyn, I. S. and Ryzhik, I. M.: *op. cit.*, 798, 810.
7. Cinco–L. H. and Meng, H.: "Pressure Transient Analysis of Wells with Finite–Conduc- tivity Vertical Fracture in Double Porosity Reservoirs," Paper SPE 18172 (1988).

III. *Withdrawal from a disc of radius ρ_f with its center at (x_w, y_w, z_w) and perpendicular to the plane $z = 0$ and at an angle ν with respect to the $x - z$ plane (isotropic medium)*

Let ρ and ρ' be the distances of points $P(x, y, z)$ and $P'(x', y', z')$ from the point $O(x_w, y_w, z_w)$, respectively. If we now temporarily consider a coordinate system with the point O as the origin, and if (ρ, θ, ϕ) and (ρ', θ', ϕ') are the coordinates of the points P and P', respectively, then, for a porous medium that is infinite in extent, the fundamental solution in spherical coordinates is given by

$$\overline{\gamma} = \frac{1}{4\pi R_D} \exp(-\sqrt{u} R_D). \tag{10}$$

Here, $R_D^2 = |PP'|^2$, and by the law of cosines

$$R_D^2 = |PP'|^2 = \rho_D^2 + \rho_D'^2 - 2\rho_D \rho_D' \cos\varphi, \tag{11}$$

where φ is the angle POP' and

$$\cos\varphi = \cos\theta\cos\theta' + \sin\theta\sin\theta'\cos(\phi - \phi'). \tag{12}$$

Using the addition theorem of Bessel functions (see Watson [2] and Abramowitz and Stegun [3]), we have

$$\frac{\exp(-kR_D)}{R_D} = \begin{cases} \sum_{m=0}^{\infty} \frac{(2m+1)}{\sqrt{\rho_D \rho_D'}} F_m\left(k\rho_D', k\rho_D\right) P_m(\cos\varphi); & \rho_D < \rho_D' \\ \sum_{m=0}^{\infty} \frac{(2m+1)}{\sqrt{\rho_D \rho_D'}} F_m\left(k\rho_D, k\rho_D'\right) P_m(\cos\varphi); & \rho_D > \rho_D', \end{cases} \tag{13}$$

where $P_m(x)$ represents the Legendre polynomial and

$$F_m(a, b) = K_{m+1/2}(a) I_{m+1/2}(b). \tag{14}$$

Using (13), we can write (10) as

$$\overline{\gamma} = \frac{1}{4\pi} \sum_{m=0}^{\infty} \frac{(2m+1)}{\sqrt{\rho_D \rho_D'}} F\left(\sqrt{u}\rho_D', \sqrt{u}\rho_D\right) P_m(\cos\varphi), \tag{15}$$

where, for notational simplicity, we have defined

$$F(a, b) = \begin{cases} F_m(a, b) & b < a \\ F_m(b, a) & b > a. \end{cases} \tag{16}$$

If we now use the method of images and let $\overline{\gamma}_n$ denote the solution for the nth image in an infinite array of sources, then the fundamental solution for a slab reservoir with impermeable boundaries at $z_D = 0$ and $z_D = z_{eD}$ can be written as (see Appendix)

$$\overline{\gamma} = \sum_{n=-\infty}^{+\infty} \overline{\gamma}_n \left(\rho_{Dn}', \rho_{Dn}, \varphi_n\right). \tag{17}$$

ds Through Porous Media, McGraw–

of Heat in Solids, Oxford University

and φ, respectively, and then recasting (17),

$$)\left[\frac{F(\sqrt{u}\rho'_D, \sqrt{u}\rho_{Dn1})}{\sqrt{\rho_{Dn1}}} P_m(\cos\varphi_{n1}) \right. \tag{18}$$

$$\left. \cos\varphi_{n2})\right],$$

$$_{wD} - 2nz_{eD})^2 + \rho_D^2 \sin^2\theta, \tag{19}$$

$$- 2nz_{eD})^2 + \rho_D^2 \sin^2\theta, \tag{20}$$

$$_k \sin\theta'_k \cos(\phi - \phi'); \; k = n1 \text{ or } n2 \tag{21}$$

$$\frac{\cos\theta + 2z_{wD} - 2nz_{eD}}{\rho_{Dn1}}, \tag{22}$$

$$\frac{\rho_D \cos\theta - 2nz_{eD}}{\rho_{Dn2}}, \tag{23}$$

$$= \pi - \theta', \tag{24}$$

$$_2 = \theta'. \tag{25}$$

disc source of radius $\rho_{fD} = \rho_f/\ell$, at an angle
erpendicular to the plane $z = 0$ can now be
ing uniform flux):

$$\sum_{m=0}^{+\infty} (2m+1)(G_{n1} + G_{n2})d\rho'_D, \tag{26}$$

$$\cos\varphi_k)_{\phi'=\nu} + P_m(\cos\varphi_k)_{\phi'=\nu+\pi}]\, d\theta', \tag{27}$$

the addition theorem of associated Legendre
4], 8.814),

$$\theta' \cos(\phi - \phi')]$$

$$)\mathrm{P}_m^k(\cos\theta')\cos(\phi - \phi'); \; [0 \le \theta, \theta' \le \pi], \tag{28}$$

IV. Computational Considerations

The title of this chapter is somewhat of a misnomer. The content is intended to aid those who wish to develop efficient algorithms to compute pressure distributions in porous media. Brute-force application of the solutions given in Chapter III is computationally inefficient, and in some cases can be counterproductive. Thus, the aim here is to permit reconsideration of the solutions and recast them into forms that should aid computations. More importantly, we provide insight into the structure of the solutions. For example, we show that the solutions for pressure distributions in a parallelepiped contain the solutions for the corresponding system in a porous medium that is infinite in areal extent.

In deriving the small-s and large-s approximations, we will, for convenience, use the following approximations for u

$$u(s \to \infty) = s\omega \tag{1}$$

and

$$u(s \to 0) = s, \tag{2}$$

respectively. Here, ω is a constant. The rationale for the limiting forms of u will become evident when we consider flow in naturally-fractured porous media (see Chapter V).

4.1. The Integral $\frac{1}{s}\int_a^b K_0\left[\sqrt{u}\sqrt{(x_D-\alpha)^2+y_D^2}\right]d\alpha$

We first consider the large-s ($t \to 0$) and the small-s ($t \to \infty$) approximations of this integral. Subsequently, we consider evaluation of the integral for $y_D = 0$. Finally, we consider the general case. Let $\overline{I}$ denote the integral.

I. *The large-s approximation*

Using the integral representation of $K_0(z)$ given by

$$K_0(z) = \frac{1}{2}\int_0^\infty \exp\left(-\xi - \frac{z^2}{4\xi}\right)\frac{d\xi}{\xi}; \ \left[\mathcal{R}e(z^2) > 0\right], \tag{1}$$

and the result

$$\int_a^b \exp\left[-\frac{(x-\alpha)^2 u}{4\xi}\right]d\alpha = \frac{\sqrt{\pi\xi}}{\sqrt{u}}\left[\operatorname{erf}\frac{(b-x)\sqrt{u}}{2\sqrt{\xi}} - \operatorname{erf}\frac{(a-x)\sqrt{u}}{2\sqrt{\xi}}\right], \tag{2}$$

we may write

$$\overline{I} = \frac{\sqrt{\pi}}{2s}\int_0^\infty \exp(-\xi)\exp\left(-\frac{uy_D^2}{4\xi}\right)\left[\operatorname{erf}\frac{(b-x_D)\sqrt{u}}{2\sqrt{\xi}} - \operatorname{erf}\frac{(a-x_D)\sqrt{u}}{2\sqrt{\xi}}\right]\frac{d\xi}{\sqrt{\xi u}}. \tag{3}$$

To obtain the large-s approximation, we replace u by $s\omega$, where ω is a constant, and thus

$$\lim_{s\to\infty}\left[\operatorname{erf}\frac{(b-x_D)\sqrt{s\omega}}{2\sqrt{\xi}}-\operatorname{erf}\frac{(a-x_D)\sqrt{s\omega}}{2\sqrt{\xi}}\right]=\beta, \tag{4}$$

where

$$\beta=\begin{cases}2 & \text{for } a<x_D<b\\ 1 & \text{for } x_D=a \text{ or } x_D=b\\ 0 & \text{for } x_D<a \text{ or } x_D>b.\end{cases} \tag{5}$$

We can thus write

$$\begin{aligned}\overline{I}&=\frac{\sqrt{\pi}\beta}{2s}\int_0^\infty \exp(-\xi)\exp\left(-\frac{\omega s y_D^2}{4\xi}\right)\frac{d\xi}{\sqrt{\omega s\xi}}\\&=\frac{\pi\beta}{2\sqrt{\omega}s^{\frac{3}{2}}}\exp\left(-|y_D|\sqrt{\omega s}\right).\end{aligned} \tag{6}$$

Application of the Inversion Theorem for the Laplace transformation yields

$$I=\beta\left[\sqrt{\pi t_D/\omega}\exp\left(-\frac{y_D^2}{4t_D/\omega}\right)-\frac{\pi}{2}|y_D|\operatorname{erfc}\left(\frac{|y_D|}{2\sqrt{t_D/\omega}}\right)\right]. \tag{7}$$

The early-time approximation for flow in homogeneous porous media is obtained by setting $\omega=1$.

II. *The small-s approximation*

We assume that $u=s$. For small values of its argument, $K_0(a\sqrt{z})$ can be written as

$$K_0(a\sqrt{z})=-\ln\left(e^\gamma a\sqrt{z}/2\right)+O(z\ln\sqrt{z}), \tag{8}$$

where γ is Euler's constant (0.5772...).

Substituting the right-hand side of (8) for $K_0(a\sqrt{z})$ in the integrand of $\overline{I}$ and evaluating the integral yields

$$\overline{I}=\frac{b-a}{2s}(-\ln s+\ln 4-2\gamma+2)+\frac{2}{s}\sigma\left(x_D,y_D,a,b\right), \tag{9}$$

where

$$\begin{aligned}\sigma(x_D,y_D,a,b)&=\frac{1}{4}\left\{(x_D-b)\ln\left[(x_D-b)^2+y_D^2\right]-(x_D-a)\ln\left[(x_D-a)^2+y_D^2\right]\right\}\\&-\frac{y_D}{2}\left[\arctan\left(\frac{x_D-a}{y_D}\right)-\arctan\left(\frac{x_D-b}{y_D}\right)\right].\end{aligned} \tag{10}$$

Application of the Inversion Theorem for the Laplace transformation yields

$$I=\frac{b-a}{2}\left[\ln\left(\frac{4t_D}{e^\gamma}\right)+2\right]+2\sigma(x_D,y_D,a,b). \tag{11}$$

III. *The case* $y_D = 0$

This case is of great importance for it can be used to obtain pressure responses for a number of systems of interest to us. Consider

$$\bar{I}(y_D = 0) = \frac{1}{s}\int_a^b K_0\left[\sqrt{u}\sqrt{(x_D-\alpha)^2}\right]d\alpha. \tag{12}$$

The Bessel function $K_0(z)$ is real and positive when $z > 0$. If $x_D \geq b$, or if $x_D \leq a$, we can write, respectively,

$$\bar{I}(y_D = 0) = \frac{1}{s\sqrt{u}}\left[\int_0^{\sqrt{u}(x_D-a)} K_0(\alpha)d\alpha - \int_0^{\sqrt{u}(x_D-b)} K_0(\alpha)d\alpha\right], \tag{13a}$$

and

$$\bar{I}(y_D = 0) = \frac{1}{s\sqrt{u}}\left[\int_0^{\sqrt{u}(b-x_D)} K_0(\alpha)d\alpha - \int_0^{\sqrt{u}(a-x_D)} K_0(\alpha)d\alpha\right]. \tag{13b}$$

If $a \leq x_D \leq b$, then we first write

$$\bar{I}(y_D = 0) = \frac{1}{s}\left\{\int_a^{x_D} K_0\left[\sqrt{u}\sqrt{(x_D-\alpha)^2}\right]d\alpha + \int_{x_D}^b K_0\left[\sqrt{u}\sqrt{(\alpha-x_D)^2}\right]d\alpha\right\}, \tag{14}$$

and, using (13a) and (13b), we have

$$\bar{I}(y_D = 0) = \frac{1}{s\sqrt{u}}\left[\int_0^{\sqrt{u}(x_D-a)} K_0(\alpha)d\alpha + \int_0^{\sqrt{u}(b-x_D)} K_0(\alpha)d\alpha\right]. \tag{15}$$

The following power-series expansion for $\int_0^x K_0(\xi)d\xi$ given by Abramowitz and Stegun [1] can be used to compute the right-hand sides of (13) and (15):

$$\begin{aligned}\int_0^x K_0(\xi)d\xi = &-\left(\ln\frac{x}{2}+\gamma\right)x\sum_{k=0}^{\infty}\frac{(x/2)^{2k}}{(k!)^2(2k+1)} \\ &+ x\sum_{k=0}^{\infty}\frac{(x/2)^{2k}}{(k!)^2(2k+1)^2} + x\sum_{k=1}^{\infty}\frac{(x/2)^{2k}}{(k!)^2(2k+1)}\sum_{n=1}^{k}\frac{1}{n},\end{aligned} \tag{16}$$

where γ is Euler's constant ($\gamma = 0.5772\ldots$). As $x \to \infty$, however, the following relation is known:

$$\int_0^{\infty} K_0(\xi)d\xi = \frac{\pi}{2}. \tag{17}$$

For $x \geq 20$, the right-hand side of (16) approaches $\pi/2$ for all practical purposes. Polynomial approximations to compute the integral in (16) are available in Luke [2].

Another alternative is to use the recurrence relation for the integrals of Bessel functions given in Abramowitz and Stegun [3],

$$\int_0^x Z_0(t)dt = xZ_0(x) + \frac{\pi x}{2}\left[-\mathbf{L}_0(x)Z_1(x) + \mathbf{L}_1(x)Z_0(x)\right], \tag{18}$$

where $\mathbf{L}_0(x)$ and $\mathbf{L}_1(x)$ are Struve functions, and

$$Z_\nu(x) = AI_\nu(x) + Be^{i\nu\pi}K_\nu(x); \quad \nu = 0, 1. \tag{19}$$

In (19), A and B are constants. Using (18) for the integral in (16), the following expression is obtained (see also Kuchuk [4] and Gradshteyn and Ryzhik [5], 6.561-4):

$$\int_0^x K_0(\xi)d\xi = xK_0(x) + \frac{\pi}{2}x\left[K_0(x)\mathbf{L}_1(x) + K_1(x)\mathbf{L}_0(x)\right]. \tag{20}$$

In this context, we should note that the integral $\int_0^x K_0(\xi)d\xi$ can be written by using (17) as follows:

$$\int_0^x K_0(\xi)d\xi = \frac{\pi}{2} - Ki_1(x), \tag{21}$$

where

$$Ki_1(x) = \int_x^\infty K_0(\xi)d\xi. \tag{22}$$

By using the integral representation for $K_\nu(z)$ given by

$$K_\nu(z) = \int_0^\infty \cosh(\nu\xi)\exp(-z\cosh\xi)d\xi; \ |\arg z| < \frac{\pi}{2}, \tag{23}$$

we can also write

$$\begin{aligned} Ki_1(x) &= \int_0^\infty dz \int_x^\infty \exp(-\xi\cosh z)\, d\xi \\ &= \int_0^\infty \frac{\exp(-x\cosh\xi)}{\cosh\xi}d\xi. \end{aligned} \tag{24}$$

In passing, we note that the application of the Inversion Theorem for the Laplace transformation to (12), assuming $u = s$, yields

$$\begin{aligned} I(x_D, y_D = 0) &= (\pi t_D)^{\frac{1}{2}}\left[\operatorname{erf}\frac{b - x_D}{2\sqrt{t_D}} - \operatorname{erf}\frac{a - x_D}{2\sqrt{t_D}}\right] \\ &- \frac{(b - x_D)}{2}\operatorname{Ei}\left[-\frac{(b - x_D)^2}{4t_D}\right] + \frac{(a - x_D)}{2}\operatorname{Ei}\left[-\frac{(a - x_D)^2}{4t_D}\right]. \end{aligned} \tag{25}$$

Here, $-\operatorname{Ei}(-x)$ is the exponential integral, $\int_x^\infty \exp(-\xi)d\xi/\xi$. (25) was first given by Gringarten, Ramey, and Raghavan [6] for $a = -1$ and $b = +1$.

IV. *The general case*

Here, we present alternate expressions for the integral $\overline{I}$. Let us first write $\overline{I}$ in a form similar to that given in (13) and (15). If $x_D \geq b$, then

$$\overline{I} = \frac{1}{s\sqrt{u}}\left[\int_0^{\sqrt{u}(x_D-a)} K_0\left(\sqrt{\xi^2+uy_D^2}\right)d\xi - \int_0^{\sqrt{u}(x_D-b)} K_0\left(\sqrt{\xi^2+uy_D^2}\right)d\xi\right]. \tag{26}$$

If $x_D \leq a$, then

$$\overline{I} = \frac{1}{s\sqrt{u}}\left[\int_0^{\sqrt{u}(b-x_D)} K_0\left(\sqrt{\xi^2+uy_D^2}\right)d\xi - \int_0^{\sqrt{u}(a-x_D)} K_0\left(\sqrt{\xi^2+uy_D^2}\right)d\xi\right], \tag{27}$$

and, if $a \leq x_D \leq b$, we obtain

$$\overline{I} = \frac{1}{s\sqrt{u}}\left[\int_0^{\sqrt{u}(x_D-a)} K_0\left(\sqrt{\xi^2+uy_D^2}\right)d\xi + \int_0^{\sqrt{u}(b-x_D)} K_0\left(\sqrt{\xi^2+uy_D^2}\right)d\xi\right]. \tag{28}$$

It is possible to express the integrals in (26)–(28) in alternate forms. For example, using (see Stakgold [7])

$$K_0\left(\sqrt{\xi^2+a^2}\right) = \frac{1}{2}\int_{-\infty}^{+\infty} e^{\mathrm{i}at}\frac{\exp\left(-\xi\sqrt{1+t^2}\right)}{\sqrt{1+t^2}}dt, \tag{29}$$

we may write

$$\begin{aligned}\int_0^x K_0\left(\sqrt{\xi^2+a^2}\right)d\xi &= \frac{1}{2}\int_0^x\int_{-\infty}^{+\infty} e^{\mathrm{i}at}\frac{\exp\left(-\xi\sqrt{1+t^2}\right)}{\sqrt{1+t^2}}dtd\xi \\ &= \frac{1}{2}\int_{-\infty}^{+\infty}\frac{e^{\mathrm{i}at}}{1+t^2}dt - \frac{1}{2}\int_{-\infty}^{+\infty} e^{\mathrm{i}at}\frac{\exp\left(-x\sqrt{1+t^2}\right)}{1+t^2}dt, \\ &= \frac{\pi}{2}\exp(-|a|) - \int_x^{\infty} K_0\left(\sqrt{\xi^2+a^2}\right)d\xi,\end{aligned} \tag{30}$$

where we have used (29) to write

$$\int_x^{\infty} K_0\left(\sqrt{\xi^2+a^2}\right)d\xi = \frac{1}{2}\int_{-\infty}^{+\infty} e^{\mathrm{i}at}\frac{\exp\left(-x\sqrt{1+t^2}\right)}{1+t^2}dt. \tag{31}$$

We can also use the integral representation of $K_0(z)$ given in (1) to obtain

$$\int_0^x K_0\left(\sqrt{\xi^2+a^2}\right)d\xi = \frac{\sqrt{\pi}}{2}\int_0^{\infty}\exp\left(-\xi-\frac{a^2}{4\xi}\right)\mathrm{erf}\left(\frac{x}{2\sqrt{\xi}}\right)\frac{d\xi}{\sqrt{\xi}}, \tag{32}$$

or we can use (see Abramowitz and Stegun [8])

$$K_0(az) = \int_0^{\infty}\frac{\cos(at)}{\sqrt{t^2+z^2}}dt;\ a>0, |\arg z| < \frac{\pi}{2} \tag{33}$$

to write

$$\int_0^x K_0\left(\sqrt{\xi^2+a^2}\right) d\xi = \int_0^\infty \cos t \ln\left(\frac{x+\sqrt{x^2+a^2+t^2}}{\sqrt{a^2+t^2}}\right) dt. \tag{34}$$

It does not appear possible to write (32) or (34) in simpler forms in terms of tabulated or known functions. The Inversion Integral for (32), however, has been available for quite some time; see Uraiet, Raghavan, and Thomas [9]. In passing, we note from (30) that

$$\int_0^\infty K_0\left(\sqrt{\xi^2+a^2}\right) d\xi = \frac{\pi}{2}\exp(-|a|), \tag{35}$$

which is a result given in Gradshteyn and Ryzhik [10], 6.596-3.

It is also useful to express the integral $\overline{I}$ in polar coordinates. Let the polar coordinates of the points (x_D, y_D) and (x'_D, y'_D) be given by (r_D, θ) and (r'_D, θ'), respectively. Consider

$$\overline{\overline{I}} = \frac{1}{s}\int_{a\cos\theta'}^{b\cos\theta'} K_0\left[\sqrt{u}\sqrt{(x_D-x'_D)^2+(y_D-x'_D\tan\theta')^2}\right] dx'_D. \tag{36}$$

Note that $\overline{\overline{I}}(\theta'=0)=\overline{I}$. We may also write the integral $\overline{\overline{I}}$ in polar coordinates as follows: Let $0 \le \theta' \le \pi/2$. If $0 \le a < b$, then

$$\overline{\overline{I}} = \frac{1}{s}\int_a^b K_0\left[\sqrt{u}\sqrt{r_D^2+r_D'^2-2r_Dr'_D\cos(\theta-\theta')}\right] dr'_D. \tag{37}$$

If $a < b \le 0$, then

$$\overline{\overline{I}} = \frac{1}{s}\int_{|b|}^{|a|} K_0\left[\sqrt{u}\sqrt{r_D^2+r_D'^2-2r_Dr'_D\cos(\theta-\theta'-\pi)}\right] dr'_D. \tag{38}$$

If $a \le 0 \le b$, then

$$\begin{aligned}\overline{\overline{I}} =& \frac{1}{s}\left\{\int_{a\cos\theta'}^{0} K_0\left[\sqrt{u}\sqrt{(x_D-x'_D)^2+(y_D-x'_D\tan\theta')^2}\right] dx'_D\right.\\ &\left.+\int_0^{b\cos\theta'} K_0\left[\sqrt{u}\sqrt{(x_D-x'_D)^2+(y_D-x'_D\tan\theta')^2}\right] dx'_D\right\},\end{aligned} \tag{39}$$

and using (37) and (38), we obtain

$$\begin{aligned}\overline{\overline{I}} =& \frac{1}{s}\left\{\int_0^{|a|} K_0\left[\sqrt{u}\sqrt{r_D^2+r_D'^2-2r_Dr'_D\cos(\theta-\theta'-\pi)}\right] dr'_D\right.\\ &\left.+\int_0^{b} K_0\left[\sqrt{u}\sqrt{r_D^2+r_D'^2-2r_Dr'_D\cos(\theta-\theta')}\right] dr'_D\right\}.\end{aligned} \tag{40}$$

Integrals appearing in the right-hand sides of (37), (38), and (40) are of the form

$$\bar{\bar{I}}_1 = \frac{1}{s}\int_c^d K_0\left[\sqrt{u}\sqrt{r_D^2 + r_D'^2 - 2r_D r_D' \cos(\alpha - \beta)}\right] dr_D'; \quad 0 \le c \le d. \qquad (41)$$

Using the addition theorem for Bessel functions given in 2.3(5), we may write

$$\bar{\bar{I}}_1 = \sum_{k=-\infty}^{+\infty} \cos k(\alpha - \beta) \int_c^d F_k dr_D', \qquad (42)$$

where $\int_c^d F_k dr_D'$ is given by

$$\begin{cases} \int_c^d I_k(r_D'\sqrt{u})\, dr_D' K_k(r_D\sqrt{u}); \; r_D > d \\ I_k(r_D\sqrt{u}) \int_c^d K_k(r_D'\sqrt{u})\, dr_D'; \; r_D < c \\ \int_c^{r_D} I_k(r_D'\sqrt{u})\, dr_D' K_k(r_D\sqrt{u}) + I_k(r_D\sqrt{u}) \int_{r_D}^d K_k(r_D'\sqrt{u})\, dr_D'; \; c < r_D < d. \end{cases}$$

Integrals in (42) can be evaluated numerically. For $\int_0^a I_\nu(x)dx$, the following infinite-series representation is given in Gradshteyn and Ryzhik [11], 6.511-11:

$$\int_0^a I_\nu(x)dx = 2\sum_{n=0}^{\infty}(-1)^n I_{\nu+2n+1}(a); \; \mathcal{R}e\ \nu > -1. \qquad (43)$$

A representation of $\int_0^a K_\nu(x)dx$ in terms of Lommel functions is given in Luke [12].

4.2. The Series $\sum_{n=1}^{\infty} \frac{\cos n\pi z \cos n\pi z_w}{\sqrt{u+n^2\pi^2/z_{eD}^2+a^2}} \exp\left(-\sqrt{u+n^2\pi^2/z_{eD}^2+a^2}\ y_D\right); y_D \ge 0$

Here, we first consider a formulation for this series when u is large. We then examine the series when u is small.

Let u denote the Laplace transform variable with respect to τ, then

$$\sum_{n=1}^{\infty} \frac{\cos n\pi x}{\sqrt{u + \frac{n^2\pi^2}{z_{eD}^2} + a^2}} \exp\left(-\sqrt{u + \frac{n^2\pi^2}{z_{eD}^2} + a^2 y_D}\right) = \mathcal{L}\{F\}, \qquad (1)$$

where $\mathcal{L}$ denotes the Laplace transform operator and F is given by

$$F = \frac{1}{\sqrt{\pi\tau}} \exp\left(-\frac{y_D^2}{4\tau}\right) \exp\left(-a^2\tau\right) \sum_{n=1}^{\infty} \cos n\pi x \exp\left(-\frac{n^2\pi^2}{z_{eD}^2}\tau\right). \qquad (2)$$

Using Poisson's summation formula given by

$$\sum_{n=-\infty}^{+\infty} \exp\left[-\frac{(\xi - 2n\xi_e)^2}{4\tau}\right] = \frac{\sqrt{\pi\tau}}{\xi_e}\left[1 + 2\sum_{n=1}^{\infty} \exp\left(-\frac{n^2\pi^2}{\xi_e^2}\tau\right) \cos n\pi \frac{\xi}{\xi_e}\right], \qquad (3)$$

the right-hand side of (2) can be written as

$$F = \frac{z_{eD} \exp\left(-a^2\tau\right)}{2\pi\tau} \sum_{n=-\infty}^{+\infty} \exp\left[-\frac{(x-2n)^2 z_{eD}^2 + y_D^2}{4\tau}\right] - \frac{\exp\left(-\frac{y_D^2}{4\tau}\right) \exp\left(-a^2\tau\right)}{2\sqrt{\pi\tau}}. \quad (4)$$

The Laplace transform of (4) is

$$\mathcal{L}\{F\} = \frac{z_{eD}}{\pi} \sum_{n=-\infty}^{+\infty} K_0\left[\sqrt{(x-2n)^2 z_{eD}^2 + y_D^2}\sqrt{u+a^2}\right] - \frac{\exp\left(-\sqrt{u+a^2}\ y_D\right)}{2\sqrt{u+a^2}}. \quad (5)$$

Substituting the right-hand side of (5) for $\mathcal{L}\{F\}$ in (1), we obtain the following relation:

$$\begin{aligned} &\sum_{n=1}^{\infty} \frac{\cos n\pi x}{\sqrt{u + \frac{n^2\pi^2}{z_{eD}^2} + a^2}} \exp\left[-\sqrt{u + \frac{n^2\pi^2}{z_{eD}^2} + a^2}\ y_D\right] \\ &= \frac{z_{eD}}{\pi} \sum_{n=-\infty}^{+\infty} K_0\left[\sqrt{(x-2n)^2 z_{eD}^2 + y_D^2}\sqrt{u+a^2}\right] \\ &- \frac{\exp\left(-\sqrt{u+a^2}\ y_D\right)}{2\sqrt{u+a^2}}. \end{aligned} \quad (6)$$

Using the relation given by (6), we finally obtain

$$\begin{aligned} &\sum_{n=1}^{\infty} \frac{\cos n\pi z \cos n\pi z_w}{\widetilde{\epsilon}_n} \exp\left(-\widetilde{\epsilon}_n y_D\right) \\ &= \frac{z_{eD}}{2\pi} \sum_{n=-\infty}^{+\infty} \left\{ K_0\left[\sqrt{(z - z_w - 2n)^2 z_{eD}^2 + y_D^2}\sqrt{u+a^2}\right]\right. \\ &\left. + K_0\left[\sqrt{(z + z_w - 2n)^2 z_{eD}^2 + y_D^2}\sqrt{u+a^2}\right]\right\} - \frac{\exp\left(-\sqrt{u+a^2}\ y_D\right)}{2\sqrt{u+a^2}}. \end{aligned} \quad (7)$$

Here, $\widetilde{\epsilon}_n = \sqrt{u + n^2\pi^2/z_{eD}^2 + a^2}$. As already noted, (7) is useful if u is large.

One other option, which may be rather convenient from a practical viewpoint, is to note the result given in Gradshteyn and Ryzhik [13], 1.448-2,

$$\sum_{k=1}^{\infty} \frac{\cos kx}{k} \exp(-ky) = \frac{1}{2} \ln \frac{1}{1 - 2\exp(-y)\cos x + \exp(-2y)}; \; y \geq 0, 0 < x \leq 2\pi. \quad (8)$$

Thus, we may write

$$\begin{aligned} \sum_{n=1}^{\infty} \frac{\cos n\pi z}{\widetilde{\epsilon}_n} \exp\left(-\widetilde{\epsilon}_n y_D\right) &= \sum_{n=1}^{\infty} \cos n\pi z \left[\frac{\exp\left(-\widetilde{\epsilon}_n y_D\right)}{\widetilde{\epsilon}_n} - \frac{\exp\left(-n\pi y_D/z_{eD}\right)}{n\pi/z_{eD}}\right] \\ &+ \frac{z_{eD}}{2\pi} \ln \frac{1}{1 - 2\exp\left(-\pi y_D/z_{eD}\right)\cos\pi z + \exp\left(-2\pi y_D/z_{eD}\right)}. \end{aligned} \quad (9)$$

This formula is useful when $(u + a^2) << n^2\pi^2/z_{eD}^2$.

4.3. The Series $\sum_{n=1}^{\infty} \frac{\sin n\pi z \sin n\pi z_w}{\sqrt{u+n^2\pi^2/z_{eD}^2+a^2}} \exp\left(-\sqrt{u + n^2\pi^2/z_{eD}^2 + a^2}\; y_D\right);\; y_D \geq 0$

Using ideas similar to those used in §4.2, for large values of u we obtain

$$\sum_{n=1}^{\infty} \frac{\sin n\pi z \sin n\pi z_w}{\sqrt{u + \frac{n^2\pi^2}{z_{eD}^2} + a^2}} \exp\left[-\sqrt{u + \frac{n^2\pi^2}{z_{eD}^2} + a^2 y_D}\right]$$
$$= \frac{z_{eD}}{2\pi} \sum_{n=-\infty}^{+\infty} \left\{ K_0 \left[\sqrt{(z - z_w - 2n)^2 z_{eD}^2 + y_D^2}\sqrt{u + a^2}\right] \right. \tag{1}$$
$$\left. - K_0 \left[\sqrt{(z + z_w - 2n)^2 z_{eD}^2 + y_D^2}\sqrt{u + a^2}\right] \right\}.$$

If $(u + a^2) << n^2\pi^2/z_{eD}^2$, then the formula in 4.2(9) should be useful.

4.4. The Series $\sum_{n=1}^{\infty} \frac{\cos(2n-1)\frac{\pi}{2} z \cos(2n-1)\frac{\pi}{2} z_w}{\sqrt{u+(2n-1)^2\pi^2/(4z_{eD}^2)+a^2}}$ $\exp\left[-\sqrt{u + (2n-1)^2 \pi^2/(4z_{eD}^2) + a^2}\; y_D\right];\; y_D \geq 0$

By arguments similar to those in §4.2, we can show that

$$\sum_{n=1}^{\infty} \frac{\cos(2n-1)\frac{\pi}{2}x}{\sqrt{u + \frac{(2n-1)^2\pi^2}{4z_{eD}^2} + a^2}} \exp\left[-\sqrt{u + \frac{(2n-1)^2\pi^2}{4z_{eD}^2} + a^2}\; y_D\right]$$
$$= \frac{z_{eD}}{\pi} \sum_{n=-\infty}^{+\infty} (-1)^n K_0 \left[\sqrt{(x - 2n)^2 z_{eD}^2 + y_D^2}\sqrt{u + a^2}\right]. \tag{1}$$

Using (1), we obtain the following formula for large u:

$$\sum_{n=1}^{\infty} \frac{\cos(2n-1)\frac{\pi}{2} z \cos(2n-1)\frac{\pi}{2} z_w}{\sqrt{u + \frac{(2n-1)^2\pi^2}{4z_{eD}^2} + a^2}} \exp\left[-\sqrt{u + \frac{(2n-1)^2\pi^2}{4z_{eD}^2} + a^2}\; y_D\right]$$
$$= \frac{z_{eD}}{2\pi} \sum_{n=-\infty}^{+\infty} (-1)^n \left\{ K_0 \left[\sqrt{(z - z_w - 2n)^2 z_{eD}^2 + y_D^2}\sqrt{u + a^2}\right] \right. \tag{2}$$
$$\left. + K_0 \left[\sqrt{(z + z_w - 2n)^2 z_{eD}^2 + y_D^2}\sqrt{u + a^2}\right] \right\}.$$

To obtain a formula that is useful when $(u + a^2) << (2n-1)^2\pi^2/(2z_{eD})^2$, we note that (see Gradshteyn and Ryzhik [14]; 1.448-4)

$$\begin{aligned}\sum_{k=1}^{\infty}&\frac{\cos(2k-1)x}{(2k-1)}\exp\left[-(2k-1)y\right]\\&=\frac{1}{4}\ln\frac{1+2\exp(-y)\cos x+\exp(-2y)}{1-2\exp(-y)\cos x+\exp(-2y)};\ y\geq 0,\ 0<x\leq 2\pi,\end{aligned}\tag{3}$$

and thus obtain

$$\begin{aligned}\sum_{n=1}^{\infty}&\frac{\cos(2n-1)\frac{\pi}{2}z}{\widetilde{\epsilon}_{2n-1}}\exp\left(-\widetilde{\epsilon}_{2n-1}y_D\right)\\&=\sum_{n=1}^{\infty}\cos(2n-1)\frac{\pi}{2}z\left\{\frac{\exp\left(-\widetilde{\epsilon}_{2n-1}y_D\right)}{\widetilde{\epsilon}_{2n-1}}-\frac{\exp\left[-(2n-1)\pi y_D/(2z_{eD})\right]}{(2n-1)\pi/(2z_{eD})}\right\}\\&+\frac{z_{eD}}{2\pi}\ln\frac{1+2\exp\left[-\pi y_D/(2z_{eD})\right]\cos\frac{\pi}{2}z+\exp\left(-\pi y_D/z_{eD}\right)}{1-2\exp\left[-\pi y_D/(2z_{eD})\right]\cos\frac{\pi}{2}z+\exp\left(-\pi y_D/z_{eD}\right)},\end{aligned}\tag{4}$$

where $\widetilde{\epsilon}_{2n-1}=\sqrt{u+(2n-1)^2\pi^2/(4z_{eD}^2)+a^2}$.

4.5. The series $\frac{1}{s}\sum_{n=1}^{\infty}\cos n\pi z_D\cos n\pi z_{wD}$ $\int_a^b K_0\left[\sqrt{u+n^2\pi^2L_D^2}\ \sqrt{(x_D-\alpha)^2+y_D^2}\right]d\alpha$

We denote the sum by $\overline{F}$ and consider expressions of $\overline{F}$ for a number of circumstances.

I. *Integral representation*

Consider first the series $\sum_{k=1}^{\infty}\cos k\pi x K_0(\sqrt{k^2+\alpha^2}r)$. If we use the relation (see Gradshteyn and Ryzhik [15], 3.961-2),

$$K_0\left(\sqrt{k^2+\alpha^2}r\right)=\int_0^{\infty}\cos\alpha x\frac{\exp\left(-k\sqrt{r^2+x^2}\right)}{\sqrt{r^2+x^2}}dx;\mathcal{R}e\ k>0,\mathcal{R}e\ r>0,\alpha>0,\tag{1}$$

then we can write

$$\sum_{k=1}^{\infty}K_0\left(\sqrt{k^2+\alpha^2}r\right)\cos k\pi x=\int_0^{\infty}\frac{\cos\alpha\xi}{\sqrt{r^2+\xi^2}}\sum_{k=1}^{\infty}\exp\left(-k\sqrt{r^2+\xi^2}\right)\cos k\pi x d\xi.\tag{2}$$

Because (see Gradshteyn and Ryzhik [16], 1.461-2)

$$1+2\sum_{k=1}^{\infty}e^{-kt}\cos kx=\frac{\operatorname{sh}t}{\operatorname{ch}t-\cos x};\quad t>0,\tag{3}$$

and (see Gradshteyn and Ryzhik [17], 3.754-2)

$$\int_0^{\infty}\frac{\cos a\xi}{\sqrt{\beta^2+\xi^2}}d\xi=K_0(a\beta);\ a>0,\quad\mathcal{R}e\ \beta>0,\tag{4}$$

we obtain

$$\sum_{k=1}^{\infty} K_0\left(\sqrt{k^2+\alpha^2 r}\right)\cos k\pi x = \frac{1}{2}\int_0^{\infty} \frac{\cos\alpha\xi}{\sqrt{r^2+\xi^2}} \frac{\operatorname{sh}\sqrt{r^2+\xi^2}}{\operatorname{ch}\sqrt{r^2+\xi^2}-\cos\pi x} d\xi - \frac{1}{2}K_0(\alpha r). \tag{5}$$

Using (5), we obtain the following integral representation for $\overline{F}$:

$$\begin{aligned}\overline{F} = \frac{1}{4s}\int_a^b \Bigg\{ \int_0^{\infty} \frac{\cos\sqrt{u}\beta}{\sqrt{\tilde{r}_D^2+\beta^2}} \Bigg[&\frac{\operatorname{sh}\left(\pi L_D\sqrt{\tilde{r}_D^2+\beta^2}\right)}{\operatorname{ch}\left(\pi L_D\sqrt{\tilde{r}_D^2+\beta^2}\right)-\cos\pi(z_D-z_{wD})} \\ &+ \frac{\operatorname{sh}\left(\pi L_D\sqrt{\tilde{r}_D^2+\beta^2}\right)}{\operatorname{ch}\left(\pi L_D\sqrt{\tilde{r}_D^2+\beta^2}\right)-\cos\pi(z_D+z_{wD})}\Bigg] d\beta - 2K_0\left(\sqrt{u}\tilde{r}_D\right)\Bigg\} d\alpha,\end{aligned} \tag{6}$$

where

$$\tilde{r}_D^2 = (x_D-\alpha)^2 + y_D^2. \tag{7}$$

II. *Computational formulation*

Along the lines discussed in §4.1, we recast $\overline{F}$ into one of the following forms appropriate for computations:

$$\begin{aligned}\overline{F}(x_D \geq b) = \frac{1}{s}\sum_{n=1}^{\infty} &\frac{\cos n\pi z_D \cos n\pi z_{wD}}{\epsilon_n} \\ &\left[\int_0^{\epsilon_n(x_D-a)} K_0\left(\sqrt{\xi^2+\epsilon_n^2 y_D^2}\right)d\xi - \int_0^{\epsilon_n(x_D-b)} K_0\left(\sqrt{\xi^2+\epsilon_n^2 y_D^2}\right)d\xi\right],\end{aligned} \tag{8}$$

$$\begin{aligned}\overline{F}(x_D \leq a) = \frac{1}{s}\sum_{n=1}^{\infty} &\frac{\cos n\pi z_D \cos n\pi z_{wD}}{\epsilon_n} \\ &\left[\int_0^{\epsilon_n(b-x_D)} K_0\left(\sqrt{\xi^2+\epsilon_n^2 y_D^2}\right)d\xi - \int_0^{\epsilon_n(a-x_D)} K_0\left(\sqrt{\xi^2+\epsilon_n^2 y_D^2}\right)d\xi\right],\end{aligned} \tag{9}$$

$$\begin{aligned}\overline{F}(a \leq x_D \leq b) = \frac{1}{s}\sum_{n=1}^{\infty} &\frac{\cos n\pi z_D \cos n\pi z_{wD}}{\epsilon_n} \\ &\left[\int_0^{\epsilon_n(x_D-a)} K_0\left(\sqrt{\xi^2+\epsilon_n^2 y_D^2}\right)d\xi + \int_0^{\epsilon_n(b-x_D)} K_0\left(\sqrt{\xi^2+\epsilon_n^2 y_D^2}\right)d\xi\right],\end{aligned} \tag{10}$$

where

$$\epsilon_n = \sqrt{u + n^2\pi^2 L_D^2}. \tag{11}$$

Although it appears innocuous, the series appearing in the right-hand side of (10) converges slowly. To overcome this difficulty, using 4.1(35), we can write (10) as follows:

$$\begin{aligned}\overline{F}(a \leq x_D \leq b) = \overline{F}_1 - \frac{1}{s}\sum_{n=1}^{\infty} &\frac{\cos n\pi z_D \cos n\pi z_{wD}}{\epsilon_n} \\ &\left[\int_{\epsilon_n(x_D-a)}^{\infty} K_0\left(\sqrt{\xi^2+\epsilon_n^2 y_D^2}\right)d\xi + \int_{\epsilon_n(b-x_D)}^{\infty} K_0\left(\sqrt{\xi^2+\epsilon_n^2 y_D^2}\right)d\xi\right],\end{aligned} \tag{12}$$

where

$$\overline{F}_1 = \frac{\pi}{s} \sum_{n=1}^{\infty} \frac{\cos n\pi z_D \cos n\pi z_{wD}}{\epsilon_n} \exp(-\epsilon_n |y_D|). \tag{13}$$

The computation of the series appearing in the right-hand side of (12) does not pose difficulties. The series, $\overline{F}_1$, given in (13), however, must be recast into forms which are appropriate for computations for small and large values of s.

A. Computation of $\overline{F}_1$ for large s.

If we use 4.2(7), $\overline{F}_1$ can be written as follows:

$$\begin{aligned}\overline{F}_1 =& \frac{1}{2L_D s} \sum_{n=-\infty}^{+\infty} \left\{ K_0 \left[\sqrt{(z_D + z_{wD} - 2n)^2 / L_D^2 + y_D^2} \sqrt{u} \right] \right. \\ &+ \left. K_0 \left[\sqrt{(z_D - z_{wD} - 2n)^2 / L_D^2 + y_D^2} \sqrt{u} \right] \right\} - \frac{\pi \exp(-\sqrt{u}|y_D|)}{2s\sqrt{u}}.\end{aligned} \tag{14}$$

The series in the right-hand side of (14) converges rapidly for large values of s.

B. Computation of $\overline{F}_1$ for small s.

Noting the relation given in 4.2(9), we write $\overline{F}_1$ as follows:

$$\begin{aligned}\overline{F}_1 =& \frac{\pi}{s} \sum_{n=1}^{\infty} \cos n\pi z_D \cos n\pi z_{wD} \left[\frac{\exp(-\epsilon_n |y_D|)}{\epsilon_n} - \frac{\exp(-n\pi L_D |y_D|)}{n\pi L_D} \right] \\ &+ \frac{1}{4L_D s} \left[\ln \frac{1}{1 - 2\exp(-\pi L_D |y_D|) \cos \pi (z_D + z_{wD}) + \exp(-2\pi L_D |y_D|)} \right. \\ &+ \left. \ln \frac{1}{1 - 2\exp(-\pi L_D |y_D|) \cos \pi (z_D - z_{wD}) + \exp(-2\pi L_D |y_D|)} \right].\end{aligned} \tag{15}$$

The series given in (15) converges rapidly for small values of s.

III. *The large-s approximation*

Replacing u by ωs where ω is a constant, and using the expressions in 4.1(1), 4.1(2), and 4.1(4), we can approximate $\overline{F}$ as $s \to \infty$ by the following expression:

$$\begin{aligned}\overline{F} =& \frac{\sqrt{\pi}\beta}{2s} \sum_{n=1}^{\infty} \cos n\pi z_D \cos n\pi z_{wD} \\ & \int_0^{\infty} \exp(-\xi) \exp\left[-\frac{(\omega s + n^2\pi^2 L_D^2)\, y_D^2}{4\xi} \right] \frac{d\xi}{\sqrt{(\omega s + n^2\pi^2 L_D^2)\,\xi}},\end{aligned} \tag{16}$$

where

$$\beta = \begin{cases} 2 & \text{for } a < x_D < b \\ 1 & \text{for } x_D = a \text{ or } x_D = b \\ 0 & \text{for } x_D < a \text{ or } x_D > b. \end{cases} \tag{17}$$

Evaluating the integral in the right-hand side of (16), we obtain

$$\overline{F} = \frac{\pi\beta}{2s}\sum_{n=1}^{\infty}\frac{\cos n\pi z_D \cos n\pi z_{wD}}{\sqrt{\omega s + n^2\pi^2 L_D^2}}\exp\left(-\sqrt{\omega s + n^2\pi^2 L_D^2}|y_D|\right). \tag{18}$$

(This same result is obtained if we evaluate (8)–(10) as $s \to \infty$ and note the relation given by 4.1(35).) If we now use the relation given in 4.2(7), we can recast (18) into the following form:

$$\begin{aligned}\overline{F} =& \frac{\beta}{4L_D s}\sum_{n=-\infty}^{+\infty}\left\{K_0\left[\sqrt{(z_D + z_{wD} - 2n)^2 / L_D^2 + y_D^2}\sqrt{\omega s}\right]\right.\\ &\left.+ K_0\left[\sqrt{(z_D - z_{wD} - 2n)^2 / L_D^2 + y_D^2}\sqrt{\omega s}\right]\right\} - \frac{\pi\beta\exp(-\sqrt{\omega s}|y_D|)}{4s^{3/2}\sqrt{\omega}}.\end{aligned} \tag{19}$$

If s is large enough, the leading term of the series in (19) is

$$K_0\left[\sqrt{(z_D - z_{wD})^2 / L_D^2 + y_D^2}\sqrt{\omega s}\right]$$

and, thus, we have

$$\overline{F} = \frac{\beta}{4L_D s}K_0\left[\sqrt{(z_D - z_{wD})^2 / L_D^2 + y_D^2}\sqrt{\omega s}\right] - \frac{\pi\beta\exp(-\sqrt{\omega s}|y_D|)}{4s^{3/2}\sqrt{\omega}}. \tag{20}$$

The Inversion Theorem for the Laplace transformation, then, yields

$$\begin{aligned}F =& -\frac{\beta}{8L_D}\mathrm{Ei}\left[-\frac{(z_D - z_{wD})^2 / L_D^2 + y_D^2}{4t_D / \omega}\right]\\ &-\frac{\beta}{2}\left[\sqrt{\frac{\pi t_D}{\omega}}\exp\left(-\frac{y_D^2}{4t_D/\omega}\right) - \frac{\pi}{2}|y_D|\mathrm{erfc}\left(\frac{|y_D|}{2\sqrt{t_D/\omega}}\right)\right].\end{aligned} \tag{21}$$

IV. *The small-s approximation*

We set $u = 0$ and write (8), (9), and (12), respectively, as follows:

$$\begin{aligned}\overline{F}(x_D \geq b) =& \frac{1}{\pi L_D s}\sum_{n=1}^{\infty}\frac{\cos n\pi z_D \cos n\pi z_{wD}}{n}\\ &\left[\int_0^{n\pi L_D(x_D-a)} K_0\left(\sqrt{\xi^2 + n^2\pi^2 L_D^2 y_D^2}\right)d\xi\right.\\ &\left.-\int_0^{n\pi L_D(x_D-b)} K_0\left(\sqrt{\xi^2 + n^2\pi^2 L_D^2 y_D^2}\right)d\xi\right],\end{aligned} \tag{22}$$

$$\overline{F}(x_D \le a) = \frac{1}{\pi L_D s} \sum_{n=1}^{\infty} \frac{\cos n\pi z_D \cos n\pi z_{wD}}{n} \left[\int_0^{n\pi L_D (b-x_D)} K_0 \left(\sqrt{\xi^2 + n^2\pi^2 L_D^2 y_D^2} \right) d\xi - \int_0^{n\pi L_D (a-x_D)} K_0 \left(\sqrt{\xi^2 + n^2\pi^2 L_D^2 y_D^2} \right) d\xi \right], \tag{23}$$

and

$$\overline{F}(a \le x_D \le b) = \overline{F}_1 - \frac{1}{\pi L_D s} \sum_{n=1}^{\infty} \frac{\cos n\pi z_D \cos n\pi z_{wD}}{n} \left[\int_{n\pi L_D (x_D - a)}^{\infty} K_0 \left(\sqrt{\xi^2 + n^2\pi^2 L_D^2 y_D^2} \right) d\xi + \int_{n\pi L_D (b-x_D)}^{\infty} K_0 \left(\sqrt{\xi^2 + n^2\pi^2 L_D^2 y_D^2} \right) d\xi \right]. \tag{24}$$

In (24), $\overline{F}_1$ is given by

$$\overline{F}_1 = \frac{1}{L_D s} \sum_{n=1}^{\infty} \frac{\cos n\pi z_D \cos n\pi z_{wD}}{n} \exp\left(-n\pi L_D |y_D|\right). \tag{25}$$

If we use 4.2(9), then $\overline{F}_1$ can be written as follows:

$$\overline{F}_1 = \frac{1}{4L_D s} \left[\ln \frac{1}{1 - 2\exp\left(-\pi L_D |y_D|\right) \cos \pi \left(z_D + z_{wD}\right) + \exp\left(-2\pi L_D |y_D|\right)} + \ln \frac{1}{1 - 2\exp\left(-\pi L_D |y_D|\right) \cos \pi \left(z_D - z_{wD}\right) + \exp\left(-2\pi L_D |y_D|\right)} \right]. \tag{26}$$

In particular, for $y_D = 0$ (this case is of considerable interest for computational purposes for it enables us to compute pressure responses which can be measured in practice), we have

$$\overline{F}_1(y_D = 0) = -\frac{1}{2L_D s} \ln \left[4 \sin \frac{\pi}{2} \left(z_D + z_{wD}\right) \sin \frac{\pi}{2} \left(z_D - z_{wD}\right) \right]. \tag{27}$$

It is possible to derive another small-s approximation as shown above. Let us consider the expression

$$S = \sum_{n=1}^{\infty} \int_a^b K_0 \left(n\pi L_D \tilde{r}_D\right) d\alpha \cos n\pi \tilde{z}_D, \tag{28}$$

where $\widetilde{r}_D$ is given by (7). Using the summation formula given in Gradshteyn and Ryzhik [18], 8.526-1,

$$\sum_{k=1}^{\infty} K_0(kx)\cos kxt = \frac{1}{2}\left(\gamma + \ln\frac{x}{4\pi}\right) + \frac{\pi}{2}\left\{\frac{1}{x\sqrt{1+t^2}}\right. + \sum_{m=1}^{\infty}\left[\frac{1}{\sqrt{x^2+(2m\pi - tx)^2}} - \frac{1}{2m\pi}\right] + \left.\sum_{m=1}^{\infty}\left[\frac{1}{\sqrt{x^2+(2m\pi + tx)^2}} - \frac{1}{2m\pi}\right]\right\}; \quad x > 0, t \text{ real}, \tag{29}$$

where γ is Euler's constant ($\gamma = 0.5772\ldots$), we may formally integrate (28) to obtain

$$S = \frac{b-a}{2}\left(\ln\frac{L_D e^{\gamma}}{4} - 1\right) - \sigma(x_D, y_D, a, b) + S_L(x_{D1}, x_{D2}, \widetilde{z}_D), \tag{30}$$

where

$$S_L(x_{D1}, x_{D2}, \widetilde{z}_D) = \frac{1}{2L_D}\left\{\ln\frac{x_{D1} + \sqrt{x_{D1}^2 + y_D^2 + \widetilde{z}_D^2/L_D^2}}{x_{D2} + \sqrt{x_{D2}^2 + y_D^2 + \widetilde{z}_D^2/L_D^2}}\right. + \sum_{m=1}^{\infty}\left[\ln\frac{x_{D1} + \sqrt{x_{D1}^2 + y_D^2 + (2m+\widetilde{z}_D)^2/L_D^2}}{x_{D2} + \sqrt{x_{D2}^2 + y_D^2 + (2m+\widetilde{z}_D)^2/L_D^2}} - \frac{(b-a)L_D}{2m}\right] + \left.\sum_{m=1}^{\infty}\left[\ln\frac{x_{D1} + \sqrt{x_{D1}^2 + y_D^2 + (2m-\widetilde{z}_D)^2/L_D^2}}{x_{D2} + \sqrt{x_{D2}^2 + y_D^2 + (2m-\widetilde{z}_D)^2/L_D^2}} - \frac{(b-a)L_D}{2m}\right]\right\}, \tag{31}$$

$$x_{D1} = \begin{cases} x_D - a & \text{for } x_D \geq b \\ b - x_D & \text{for } x_D \leq a, \end{cases} \tag{32}$$

$$x_{D2} = \begin{cases} x_D - b & \text{for } x_D \geq b \\ a - x_D & \text{for } x_D \leq a, \end{cases} \tag{33}$$

and

$$\sigma(x_D, y_D, a, b) = \frac{1}{4}\left\{(x_D - b)\ln\left[(x_D-b)^2 + y_D^2\right] - (x_D - a)\ln\left[(x_D-a)^2 + y_D^2\right] - 2y_D\left(\arctan\frac{x_D - a}{y_D} - \arctan\frac{x_D - b}{y_D}\right)\right\}. \tag{34}$$

For $x_D \geq b$, we can now write

$$\overline{F}(x_D \geq b) = \frac{1}{2s}\sum_{n=1}^{\infty}\left[\cos n\pi(z_D + z_{wD}) + \cos n\pi(z_D - z_{wD})\right] \int_a^b K_0\left[n\pi L_D\sqrt{(x_D-\alpha)^2 + y_D^2}\right] d\alpha, \tag{35}$$

and, using (30), we obtain

$$\begin{aligned}\overline{F}(x_D \geq b) =& \frac{1}{2s}\left[(b-a)\left(\ln\frac{L_D e^{\gamma}}{4} - 1\right) - 2\sigma(x_D, y_D, a, b)\right.\\ &\left. + S_L(x_D - a, x_D - b, z_D + z_{wD}) + S_L(x_D - a, x_D - b, z_D - z_{wD})\right].\end{aligned} \tag{36}$$

Similarly, for $x_D \leq a$ we have

$$\begin{aligned}\overline{F}(x_D \leq a) =& \frac{1}{2s}\left[(b-a)\left(\ln\frac{L_D e^{\gamma}}{4} - 1\right) - 2\sigma(x_D, y_D, a, b)\right.\\ &\left. + S_L(b - x_D, a - x_D, z_D + z_{wD}) + S_L(b - x_D, a - x_D, z_D - z_{wD})\right].\end{aligned} \tag{37}$$

If $a \leq x_D \leq b$, then we first write

$$\begin{aligned}\overline{F}(a \leq x_D \leq b) =& \frac{1}{2s}\sum_{n=1}^{\infty}[\cos n\pi(z_D + z_{wD}) + \cos n\pi(z_D - z_{wD})]\\ &\left\{\int_a^{x_D} K_0\left[n\pi L_D\sqrt{(x_D - \alpha)^2 + y_D^2}\right]d\alpha + \int_{x_D}^{b} K_0\left[n\pi L_D\sqrt{(\alpha - x_D)^2 + y_D^2}\right]d\alpha\right\},\end{aligned} \tag{38}$$

and using (36) and (37), we obtain

$$\begin{aligned}\overline{F}(a \leq x_D \leq b) =& \frac{1}{2s}\left[(b-a)\left(\ln\frac{L_D e^{\gamma}}{4} - 1\right) - 2\sigma(x_D, y_D, a, b)\right.\\ & + S_L(x_D - a, 0, z_D + z_{wD}) + S_L(x_D - a, 0, z_D - z_{wD})\\ &\left. + S_L(b - x_D, 0, z_D + z_{wD}) + S_L(b - x_D, 0, z_D - z_{wD})\right].\end{aligned} \tag{39}$$

4.6. Flow in a domain bounded by two parallel planes that are impermeable. Some applications

Here, we consider a few examples that illustrate the utility of the approximations discussed in the previous sections. We consider flow in an isotropic reservoir ($k = k_x = k_y = k_z$) and assume that the boundaries $z = 0$ and $z = z_e$ are impermeable.

I. *A horizontal well, uniform-flux wellbore*

The center of the well is located at $(0, 0, z_w)$ and the flux distribution is assumed to be uniform. Let $\ell = L_h/2$, where L_h is the length of the well.

As discussed in §3.1, the pressure distribution is given by

$$\begin{aligned}\overline{p}_D =& \frac{1}{2s}\int_{-1}^{+1} K_0\left[\sqrt{u}\sqrt{(x_D - \alpha)^2 + y_D^2}\right]d\alpha\\ &+ \frac{1}{s}\sum_{n=1}^{\infty}\cos n\pi z_D \cos n\pi z_{wD}\int_{-1}^{+1} K_0\left[\sqrt{u + n^2\pi^2 L_D^2}\sqrt{(x_D - \alpha)^2 + y_D^2}\right]d\alpha.\end{aligned} \tag{1}$$

In (1), z_D and L_D are defined, respectively, by

$$z_D = \frac{z}{z_e} \tag{2}$$

and

$$L_D = \frac{1}{z_{eD}} = \frac{L_h}{2z_e}. \tag{3}$$

Should we now wish to determine the response along the well ($|x_D| \leq 1, y_D = 0$ and $z_D = z_{wD} + r_{wD}$, where $r_{wD} = r_w/h$ and r_w is the radius of the well), then using 4.1(15), we rewrite (1) as follows:

$$\overline{p}_D(|x_D| \leq 1, y_D = 0, z_D = z_{wD} + r_{wD}) = \frac{1}{2s\sqrt{u}}$$
$$\left[\int_0^{\sqrt{u}(1-x_D)} K_0(\xi)d\xi + \int_0^{\sqrt{u}(1+x_D)} K_0(\xi)d\xi\right] + \overline{F}(x_D, z_D, z_{wD}, L_D), \tag{4}$$

where

$$\overline{F}(x_D, z_D, z_{wD}, L_D) = \frac{1}{s}\sum_{n=1}^{\infty} \frac{\cos n\pi z_D \cos n\pi z_{wD}}{\sqrt{u + n^2\pi^2 L_D^2}}$$
$$\left[\int_0^{\sqrt{u+n^2\pi^2L_D^2}(1-x_D)} K_0(\xi)d\xi + \int_0^{\sqrt{u+n^2\pi^2L_D^2}(1+x_D)} K_0(\xi)d\xi\right]. \tag{5}$$

Methods to compute the integrals as well as the sum in the above expressions were discussed in previous sections. Solutions for $|x_D| \geq 1$ and $y_D \neq 0$ can be obtained in a similar fashion along the lines suggested in §4.1.

II. *A vertically-fractured well, uniform-flux solution*

We assume a uniform-flux wellbore and consider the center of the well to be at the $(0, 0, z_e/2)$. Let $2L_f$ be the length of the fracture, h be its height and let $\ell = L_f$. As discussed in §3.1, the pressure distribution is given by

$$\overline{p}_D(x_D, y_D) = \frac{1}{2s}\int_{-1}^{+1} K_0\left[\sqrt{u}\sqrt{(x_D - \alpha)^2 + y_D^2}\right] d\alpha. \tag{6}$$

The pressure distribution along the fracture plane ($|x_D| \leq 1, y_D = 0$) is (see §4.1 III)

$$\overline{p}_D(|x_D| \leq 1, y_D = 0) = \frac{1}{2s\sqrt{u}}\left[\int_0^{\sqrt{u}(1-x_D)} K_0(\xi)d\xi + \int_0^{\sqrt{u}(1+x_D)} K_0(\xi)d\xi\right]. \tag{7}$$

It should be noted that the Inversion Integral of (6) has been available for quite some time. Methods to compute the integrals in (6) and (7) were discussed earlier. (7) may also be used to obtain the wellbore response at wells produced via infinite-conductivity or finite-conductivity fractures.

III. *Infinite-conductivity wellbores*

We have already discussed a scheme to compute pressure responses at infinite-conductivity wellbores in §3.1 VI. Here, we consider additional schemes to compute infinite-conductivity responses. For vertically-fractured wells, Gringarten, Ramey, and Raghavan [19] showed that infinite-conductivity responses could be computed by assuming that $x_D = 0.732$ in (7). (A similar scheme was suggested by Muskat [20] for steady-state flow.) Because infinite-conductivity and uniform-flux responses are identical at early times, they suggested that this value of x_D could be used for all values of time. A refined estimate of x_D has been recently developed by Kuchuk [21]. He recommends $x_D = 0.74009714$. With regard to horizontal wells, one would expect to use an identical value of x_D. Some, however, have suggested that x_D would be a function of L_D.

Another technique to approximate infinite-conductivity wellbores is to assume that an integrated average of the uniform-flux solution will yield a good approximation of the infinite-conductivity idealization (Hantush [22], Streltsova-Adams [23], Kuchuk, Goode, Wilkinson, and Thambynayagam [24].) Wilkinson and Hammond [25] have placed this method on a firm footing. We briefly examine this option.

The infinite-conductivity response corresponding to (7) is

$$\begin{aligned}\bar{\bar{p}}_D &= \frac{1}{2}\int_{-1}^{+1} \bar{p}_D(x_D, y_D = 0, s)dx_D \\ &= \frac{1}{2s\sqrt{u}}\left[\pi - \frac{1 - Ki_2(2\sqrt{u})}{\sqrt{u}}\right],\end{aligned} \tag{8}$$

where

$$Ki_2(z) = \int_z^{\infty} Ki_1(\xi)d\xi. \tag{9}$$

Similarly, the infinite-conductivity response corresponding to (1) is

$$\bar{\bar{\bar{p}}}_D = \bar{\bar{p}}_D + \frac{1}{s}\sum_{n=1}^{\infty}\frac{\cos n\pi z_D \cos n\pi z_{wD}}{\sqrt{u + n^2\pi^2 L_D^2}}\left[\pi - \frac{1 - Ki_2\left(2\sqrt{u + n^2\pi^2 L_D^2}\right)}{\sqrt{u + n^2\pi^2 L_D^2}}\right]. \tag{10}$$

These results have been derived independently by Chen [26].

IV. *An inclined well, the small-s approximation*

Again, we assume a uniform-flux wellbore inclined at an angle θ to the vertical. Let $h_{fD} = h_f/\ell$, where h_f is the length of the well and ℓ is the reference length. The pressure distribution in this case is given by (see §3.1 IV)

$$\begin{aligned}\overline{\Delta p} = \frac{\tilde{q}\mu}{2\pi k z_{eD} s}\int_{-\frac{h_{fD}}{2}}^{+\frac{h_{fD}}{2}} &\left[K_0(\tilde{r}_D\sqrt{u}) + 2\sum_{n=1}^{\infty} K_0\left(\tilde{r}_D\sqrt{u + \frac{n^2\pi^2}{z_{eD}^2}}\right)\right. \\ &\left.\cos n\pi\frac{z_D}{z_{eD}}\cos n\pi\frac{\tilde{z}_{wD}}{z_{eD}}\right] d\alpha,\end{aligned} \tag{11}$$

where

$$\widetilde{r}_D^2 = [x_D - (x_{wD} + \alpha \sin\theta)]^2 + (y_D - y_{wD})^2 \tag{12}$$

and

$$\widetilde{z}_{wD} = z_{wD} + \alpha \cos\theta. \tag{13}$$

Replacing $K_0(z)$ by $-\ln(e^\gamma z/2)$ in (11), the small-s approximation of (11) becomes

$$\begin{aligned}\overline{\Delta p} = \frac{\widetilde{q}\mu}{2\pi k z_{eD} s} \int_{-\frac{h_{fD}}{2}}^{+\frac{h_{fD}}{2}} & \left\{ -\ln\left(\frac{e^\gamma \widetilde{r}_D \sqrt{u}}{2}\right) \right. \\ & \left. + \sum_{n=1}^{\infty} K_0\left(\frac{n\pi}{z_{eD}} \widetilde{r}_D\right) \left[\cos n\pi \frac{\widetilde{z}_{D1}}{z_{eD}} + \cos n\pi \frac{\widetilde{z}_{D2}}{z_{eD}}\right] \right\} d\alpha,\end{aligned} \tag{14}$$

where

$$\widetilde{z}_{Dj} = z_D + (-1)^{j+1}(z_{wD} + \alpha\cos\theta); \quad j = 1 \text{ or } 2. \tag{15}$$

We thus have

$$\begin{aligned}\int_{-\frac{h_{fD}}{2}}^{+\frac{h_{fD}}{2}} -\ln\left(\frac{e^\gamma \widetilde{r}_D \sqrt{u}}{2}\right) d\alpha = & -h_{fD}\left(\ln\frac{e^\gamma\sqrt{u}}{2} - 1\right) \\ & + \frac{2}{\sin\theta}\sigma\left(\widetilde{x}_D, \widetilde{y}_D, -\frac{h_{fD}}{2}\sin\theta, \frac{h_{fD}}{2}\sin\theta\right),\end{aligned} \tag{16}$$

where

$$\widetilde{x}_D = x_D - x_{wD}, \tag{17}$$

$$\widetilde{y}_D = y_D - y_{wD}, \tag{18}$$

and $\sigma(\widetilde{x}_D, \widetilde{y}_D, a, b)$ is given in 4.5(34).

Consider

$$S = \int_a^b \sum_{n=1}^{\infty} K_0\left(\frac{n\pi}{z_{eD}}\widetilde{r}_D\right) \cos n\pi \frac{\widetilde{z}_{Dj}}{z_{eD}} d\alpha. \tag{19}$$

If $\widetilde{x}_D \geq b\sin\theta$, then

$$S = \frac{1}{\sin\theta} \int_{\widetilde{x}_D - b\sin\theta}^{\widetilde{x}_D - a\sin\theta} \sum_{n=1}^{\infty} K_0\left(\frac{n\pi}{z_{eD}}\sqrt{\xi^2 + \widetilde{y}_D^2}\right) \cos n\pi \frac{\widehat{z}_{Dj} + (-1)^j \xi \text{ cotan }\theta}{z_{eD}} d\xi. \tag{20}$$

Similarly, if $\widetilde{x}_D \leq a\sin\theta$, then

$$S = \frac{1}{\sin\theta} \int_{a\sin\theta - \widetilde{x}_D}^{b\sin\theta - \widetilde{x}_D} \sum_{n=1}^{\infty} K_0\left(\frac{n\pi}{z_{eD}}\sqrt{\xi^2 + \widetilde{y}_D^2}\right) \cos n\pi \frac{\widehat{z}_{Dj} + (-1)^{j+1} \xi \text{ cotan }\theta}{z_{eD}} d\xi. \tag{21}$$

In (20) and (21)

$$\widehat{z}_{Dj} = z_D + (-1)^{j+1}(z_{wD} + \widetilde{x}_D \text{ cotan }\theta). \tag{22}$$

Using the summation formula given in 4.5(29) and evaluating the integrals, we obtain the following expression for S:

$$S = \frac{b-a}{2}\left(\ln\frac{e^{\gamma}}{4z_{eD}} - 1\right) - \frac{1}{\sin\theta}\sigma\left(\tilde{x}_D, \tilde{y}_D, a\sin\theta, b\sin\theta\right) + S_L(k, \tilde{x}_{D1}, \tilde{x}_{D2}, \hat{z}_{Dj}). \tag{23}$$

In (23), $S_L(k, \tilde{x}_{D1}, \tilde{x}_{D2}, \hat{z}_{Dj})$ is given by

$$S_L(k, \tilde{x}_{D1}, \tilde{x}_{D2}, \hat{z}_{Dj}) = \frac{z_{eD}}{2}\left[\ln\frac{\sqrt{A\tilde{x}_{D1}^2 + 2B\tilde{x}_{D1} + C} + \sqrt{A}\tilde{x}_{D1} + \frac{B}{\sqrt{A}}}{\sqrt{A\tilde{x}_{D2}^2 + 2B\tilde{x}_{D2} + C} + \sqrt{A}\tilde{x}_{D2} + \frac{B}{\sqrt{A}}}\right.$$
$$+ \sum_{m=1}^{\infty}\left(\ln\frac{\sqrt{A\tilde{x}_{D1}^2 + 2D\tilde{x}_{D1} + E} + \sqrt{A}\tilde{x}_{D1} + \frac{D}{\sqrt{A}}}{\sqrt{A\tilde{x}_{D2}^2 + 2D\tilde{x}_{D2} + E} + \sqrt{A}\tilde{x}_{D2} + \frac{D}{\sqrt{A}}} - \frac{b-a}{2mz_{eD}}\right) \tag{24}$$
$$\left. + \sum_{m=1}^{\infty}\left(\ln\frac{\sqrt{A\tilde{x}_{D1}^2 + 2F\tilde{x}_{D1} + G} + \sqrt{A}\tilde{x}_{D1} + \frac{F}{\sqrt{A}}}{\sqrt{A\tilde{x}_{D2}^2 + 2F\tilde{x}_{D2} + G} + \sqrt{A}\tilde{x}_{D2} + \frac{F}{\sqrt{A}}} - \frac{b-a}{2mz_{eD}}\right)\right],$$

where

$$k = \begin{cases} 1; & \tilde{x}_D \geq b\sin\theta \\ 2; & \tilde{x}_D \leq a\sin\theta, \end{cases} \tag{25}$$

$$\tilde{x}_{D1} = \begin{cases} \tilde{x}_D - a\sin\theta; & \tilde{x}_D \geq b\sin\theta \\ b\sin\theta - \tilde{x}_D; & \tilde{x}_D \leq a\sin\theta, \end{cases} \tag{26}$$

$$\tilde{x}_{D2} = \begin{cases} \tilde{x}_D - b\sin\theta; & \tilde{x}_D \geq b\sin\theta \\ a\sin\theta - \tilde{x}_D; & \tilde{x}_D \leq a\sin\theta, \end{cases} \tag{27}$$

$$A = 1/\sin^2\theta, \tag{28}$$

$$B = (-1)^{j+k+1}\hat{z}_{Dj}\ \text{cotan}\ \theta, \tag{29}$$

$$C = \tilde{y}_D^2 + \hat{z}_{Dj}^2, \tag{30}$$

$$D = (-1)^{j+k+1}\left(2mz_{eD} + \hat{z}_{Dj}\right)\text{cotan}\ \theta, \tag{31}$$

$$E = \tilde{y}_D^2 + \left(2mz_{eD} + \hat{z}_{Dj}\right)^2, \tag{32}$$

$$F = (-1)^{j+k}\left(2mz_{eD} - \hat{z}_{Dj}\right)\text{cotan}\ \theta, \tag{33}$$

and

$$G = \tilde{y}_D^2 + \left(2mz_{eD} - \hat{z}_{Dj}\right)^2. \tag{34}$$

If we now use (16) and (24) in (14), we can write the small-s approximation of (11) as follows:

$$\overline{\Delta p} = \frac{\tilde{q}\mu}{2\pi k z_{eD} s}\left(h_{fD}\ln\frac{1}{2z_{eD}\sqrt{u}} + S_T\right), \tag{35}$$

where S_T is given by

$$S_T\left(\tilde{x}_D \geq \frac{h_{fD}}{2}\sin\theta\right)$$
$$= S_L\left(1, \tilde{x}_D + \frac{h_{fD}}{2}\sin\theta, \tilde{x}_D - \frac{h_{fD}}{2}\sin\theta, z_D + z_{wD} + \tilde{x}_D \text{ cotan } \theta\right) \tag{36}$$
$$+ S_L\left(1, \tilde{x}_D + \frac{h_{fD}}{2}\sin\theta, \tilde{x}_D - \frac{h_{fD}}{2}\sin\theta, z_D - z_{wD} - \tilde{x}_D \text{ cotan } \theta\right),$$

$$S_T\left(\tilde{x}_D \leq -\frac{h_{fD}}{2}\sin\theta\right)$$
$$= S_L\left(2, \frac{h_{fD}}{2}\sin\theta - \tilde{x}_D, -\frac{h_{fD}}{2}\sin\theta - \tilde{x}_D, z_D + z_{wD} + \tilde{x}_D \text{ cotan } \theta\right) \tag{37}$$
$$+ S_L\left(2, \frac{h_{fD}}{2}\sin\theta - \tilde{x}_D, -\frac{h_{fD}}{2}\sin\theta - \tilde{x}_D, z_D - z_{wD} - \tilde{x}_D \text{ cotan } \theta\right),$$

and

$$S_T\left(-\frac{h_{fD}}{2}\sin\theta \leq \tilde{x}_D \leq \frac{h_{fD}}{2}\sin\theta\right)$$
$$= S_L\left(1, \tilde{x}_D + \frac{h_{fD}}{2}\sin\theta, 0, z_D + z_{wD} + \tilde{x}_D \text{ cotan } \theta\right)$$
$$+ S_L\left(1, \tilde{x}_D + \frac{h_{fD}}{2}\sin\theta, 0, z_D - z_{wD} - \tilde{x}_D \text{ cotan } \theta\right) \tag{38}$$
$$+ S_L\left(2, \frac{h_{fD}}{2}\sin\theta - \tilde{x}_D, 0, z_D + z_{wD} + \tilde{x}_D \text{ cotan } \theta\right)$$
$$+ S_L\left(2, \frac{h_{fD}}{2}\sin\theta - \tilde{x}_D, 0, z_D - z_{wD} - \tilde{x}_D \text{ cotan } \theta\right).$$

4.7. A note on the ratios $ch(\sqrt{u}\alpha)/sh(\sqrt{u}\beta)$ and $sh(\sqrt{u}\alpha)/sh(\sqrt{u}\beta)$ for large s

The expressions for the pressure distributions in a parallelepiped contain ratios of hyperbolic functions. Although we would rarely use these solutions to compute pressure distributions if we expect the boundaries to be infinite in extent in the x and y directions, the information given here is presented mainly for completeness and continuity. We assume $u = \omega s$ and consider solutions as $s \to \infty$. Consider first the term

$$\frac{ch\sqrt{u}\,[y_{eD} - (y_D + y_{wD})] + ch\sqrt{u}(y_{eD} - |y_D - y_{wD}|)}{2\sqrt{u}\; sh\sqrt{u}\; y_{eD}}.$$

Now

$$\lim_{s\to\infty}\left\{\frac{ch\sqrt{u}[y_{eD}-(y_D+y_{wD})]+ch\sqrt{u}(y_{eD}-|y_D-y_{wD}|)}{2\sqrt{u}sh\sqrt{u}y_{eD}}\right\}$$

$$=\lim_{s\to\infty}\left\{\frac{1}{2\sqrt{u}\left(1-e^{-2\sqrt{u}y_{eD}}\right)}\left[e^{-\sqrt{u}|y_D-y_{wD}|}+e^{-\sqrt{u}(y_D+y_{wD})}\right.\right. \tag{1}$$

$$\left.\left.+e^{-\sqrt{u}(2y_{eD}-|y_D-y_{wD}|)}+e^{-\sqrt{u}[2y_{eD}-(y_D+y_{wD})]}\right]\right\}.$$

If we assume that

$$\lim_{s\to\infty}\exp\left(-2\sqrt{u}y_{eD}\right)\approx 0, \tag{2}$$

then (1) can be written as

$$\lim_{s\to\infty}\left\{\frac{ch\sqrt{u}\left[y_{eD}-(y_D+y_{wD})\right]+ch\sqrt{u}(y_{eD}-|y_D-y_{wD}|)}{2\sqrt{u}sh\sqrt{u}y_{eD}}\right\}$$

$$=\frac{1}{2\sqrt{u}}\left\{e^{-\sqrt{u}|y_D-y_{wD}|}+e^{-\sqrt{u}(y_D+y_{wD})}\right. \tag{3}$$

$$\left.+e^{-\sqrt{u}(2y_{eD}-|y_D-y_{wD}|)}+e^{-\sqrt{u}[2y_{eD}-(y_D+y_{wD})]}\right\}.$$

Similarly, assuming that the condition given by (2) holds, we can write

$$\lim_{s\to\infty}\left\{\frac{ch\sqrt{u}\left[y_{eD}-(y_D+y_{wD})\right]-ch\sqrt{u}(y_{eD}-|y_D-y_{wD}|)}{2\sqrt{u}sh\sqrt{u}y_{eD}}\right\}$$

$$=\frac{1}{2\sqrt{u}}\left\{e^{-\sqrt{u}|y_D-y_{wD}|}-e^{-\sqrt{u}(y_D+y_{wD})}\right. \tag{4}$$

$$\left.+e^{-\sqrt{u}(2y_{eD}-|y_D-y_{wD}|)}-e^{-\sqrt{u}[2y_{eD}-(y_D+y_{wD})]}\right\},$$

and

$$\lim_{s\to\infty}\left\{\frac{sh\sqrt{u}\left[y_{eD}-(y_D+y_{wD})\right]+sh\sqrt{u}(y_{eD}-|y_D-y_{wD}|)}{2\sqrt{u}sh\sqrt{u}y_{eD}}\right\}$$

$$=\frac{1}{2\sqrt{u}}\left\{e^{-\sqrt{u}|y_D-y_{wD}|}+e^{-\sqrt{u}(y_D+y_{wD})}\right. \tag{5}$$

$$\left.-e^{-\sqrt{u}(2y_{eD}-|y_D-y_{wD}|)}-e^{-\sqrt{u}[2y_{eD}-(y_D+y_{wD})]}\right\}.$$

If we also assume that the following conditions hold

$$\exp\left[-\sqrt{u}(y_D+y_{wD})\right]\approx 0, \tag{6}$$

$$\exp\left\{-\sqrt{u}(2y_{eD}-|y_D-y_{wD}|)\right\}\approx 0, \tag{7}$$

$$\exp\left\{-\sqrt{u}\left[2y_{eD}-(y_D+y_{wD})\right]\right\}\approx 0, \tag{8}$$

and let HF denote the ratios of the hyperbolic functions on the left-hand sides of (3), (4), and (5), we can write

$$\lim_{s\to\infty}\{HF\}=\frac{1}{2\sqrt{u}}\exp\left(-\sqrt{u}|y_D-y_{wD}|\right). \tag{9}$$

(9) can be used to replace the ratios of the hyperbolic functions in the solutions given in §3.3 when s is large. For example, consider the solution in 3.3(2) (a vertically-fractured well in a reservoir wherein all boundaries are impermeable)

$$\bar{p}_D(x_D, y_D) = \frac{\pi}{x_{eD}s}\left[\frac{ch\sqrt{u}(y_{eD} - |y_D - y_{wD}|) + ch\sqrt{u}[y_{eD} - (y_D + y_{wD})]}{\sqrt{u}\; sh\sqrt{u}y_{eD}}\right.$$

$$+ \frac{2x_{eD}}{\pi}\sum_{k=1}^{\infty}\frac{1}{k}\sin\ k\pi\frac{1}{x_{eD}}\cos\ k\pi\frac{x_{wD}}{x_{eD}}\cos\ k\pi\frac{x_D}{x_{eD}} \tag{10}$$

$$\left.\frac{ch\sqrt{u+\frac{k^2\pi^2}{x_{eD}^2}}(y_{eD} - |y_D - y_{wD}|) + ch\sqrt{u+\frac{k^2\pi^2}{x_{eD}^2}}[y_{eD} - (y_D + y_{wD})]}{\sqrt{u+\frac{k^2\pi^2}{x_{eD}^2}}\,sh\sqrt{u+\frac{k^2\pi^2}{x_{eD}^2}}\,y_{eD}}\right].$$

Using (9), and substituting $s\omega$ for u, we may write (10) as

$$\lim_{s\to\infty}\left\{\frac{sx_{eD}}{2\pi}\bar{p}_D\right\} = \frac{1}{2\sqrt{s\omega}}\exp\left(-\sqrt{s\omega}|y_D - y_{wD}|\right)$$

$$+ \frac{2x_{eD}}{\pi}\sum_{k=1}^{\infty}\frac{1}{k}\sin k\pi\frac{1}{x_{eD}}\cos k\pi\frac{x_{wD}}{x_{eD}}\cos k\pi\frac{x_D}{x_{eD}} \tag{11}$$

$$\frac{1}{2\sqrt{s\omega+\frac{k^2\pi^2}{x_{eD}^2}}}\exp\left(-\sqrt{s\omega+\frac{k^2\pi^2}{x_{eD}^2}}|y_D - y_{wD}|\right).$$

If we let $s\omega$ denote the Laplace transform variable with respect to ξ, and $\mathcal{L}$ denote the Laplace transform operator, then, inverting the Laplace transformation of the right-hand side of (11) with respect to $s\omega$, we may write

$$\lim_{s\to\infty}\left\{\frac{sx_{eD}}{2\pi}\bar{p}_D\right\} = \mathcal{L}\left\{\frac{1}{2\sqrt{\pi\xi}}\exp\left[-\frac{(y_D - y_{wD})^2}{4\xi}\right]\right.$$

$$\left[1 + \frac{2x_{eD}}{\pi}\sum_{k=1}^{\infty}\frac{1}{k}\sin k\pi\frac{1}{x_{eD}}\cos k\pi\frac{x_{wD}}{x_{eD}}\cos k\pi\frac{x_D}{x_{eD}}\right. \tag{12}$$

$$\left.\left.\exp\left(-\frac{k^2\pi^2\xi}{x_{eD}^2}\right)\right]\right\}.$$

As $\xi \to 0$, the terms within the square brackets in (12) yield $\beta x_{eD}/4$ (see Gringarten and Ramey [27]), where

$$\beta = \begin{cases} 2 & \text{for } |x_D - x_{wD}| < 1, \\ 1 & \text{for } |x_D - x_{wD}| = 1, \\ 0 & \text{for } |x_D - x_{wD}| > 1. \end{cases} \tag{13}$$

Therefore, (12) may be written as

$$\lim_{s\to\infty}\left\{\frac{sx_{eD}}{2\pi}\bar{p}_D\right\} = \mathcal{L}\left\{\frac{\beta x_{eD}}{8\sqrt{\pi\xi}}\exp\left[-\frac{(y_D - y_{wD})^2}{4\xi}\right]\right\}. \tag{14}$$

Taking the Laplace transformation of the function inside the curly braces on the right-hand side of (14) with respect to ξ, we obtain the following expression:

$$\lim_{s\to\infty} \bar{p}_D = \frac{\pi\beta}{4s\sqrt{s\omega}} \exp\left(-\sqrt{s\omega}|y_D - y_{wD}|\right). \tag{15}$$

Thus, the short-time approximation of the vertically-fractured-well solution is

$$p_D = \frac{\beta}{2}\left\{\sqrt{\pi t_D/\omega}\exp\left[-\frac{(y_D - y_{wD})^2}{4t_D/\omega}\right] - \frac{\pi}{2}|y_D - y_{wD}|\mathrm{erfc}\left(\frac{|y_D - y_{wD}|}{2\sqrt{t_D/\omega}}\right)\right\}. \tag{16}$$

Not unexpectedly, the right-hand side of (16) is one-half the short-time approximation given by 4.1(7).

4.8. Flow in cylindrical porous media. Some Applications

Here, we address issues regarding the computation of pressure responses in cylindrical porous media. We consider the pressure distribution solution, given in 3.2(7), for a vertically-fractured well located centrally in a cylindrical reservoir with all boundaries impermeable. If we write

$$\frac{1}{(-1)^n}\frac{K_n'(\sqrt{u}r_{eD})}{I_n'(\sqrt{u}r_{eD})} = \frac{K_0'(\sqrt{u}r_{eD})}{I_0'(\sqrt{u}r_{eD})}, \tag{1}$$

and note that; see Watson [28]

$$I_0(R_D\sqrt{u}) = \sum_{n=-\infty}^{+\infty}(-1)^n I_n(r_D\sqrt{u})I_n(r_D'\sqrt{u})\cos n(\theta - \theta'), \tag{2}$$

then we can write 3.2(7) as

$$\begin{aligned}\bar{p}_D = \bar{p}_{Di} + \frac{1}{2s}\frac{K_1(r_{eD}\sqrt{u})}{I_1(r_{eD}\sqrt{u})}\int_0^1\Big\{I_0\left[\sqrt{u}\sqrt{r_D^2 + r_D'^2 - 2r_Dr_D'\cos(\theta-\alpha)}\right]\\ + I_0\left[\sqrt{u}\sqrt{r_D^2 + r_D'^2 - 2r_Dr_D'\cos(\theta-\alpha-\pi)}\right]\Big\}dr_D'.\end{aligned} \tag{3}$$

Because (see §4.1 IV)

$$\begin{aligned}\int_{-\cos\theta'}^{+\cos\theta'} Z\left(\sqrt{u}\widehat{R}_D\right)d\xi = \int_0^1\Big\{Z\left[\sqrt{u}\sqrt{r_D^2 + r_D'^2 - 2r_Dr_D'\cos(\theta-\theta')}\right]\\ + Z\left[\sqrt{u}\sqrt{r_D^2 + r_D'^2 - 2r_Dr_D'\cos(\theta-\theta'-\pi)}\right]\Big\}dr_D',\end{aligned} \tag{4}$$

where

$$\widehat{R}_D^2 = (x_D - x_{wD} - \xi)^2 + (y_D - y_{wD} - \xi\tan\alpha)^2, \tag{5}$$

we have

$$\bar{p}_D = \frac{1}{2s}\int_{-\cos\alpha}^{+\cos\alpha}\left[K_0\left(\sqrt{u}\widehat{R}_D\right) + I_0\left(\sqrt{u}\widehat{R}_D\right)\frac{K_1\left(\sqrt{u}r_{eD}\right)}{I_1\left(\sqrt{u}r_{eD}\right)}\right]d\xi, \tag{6}$$

Although (1) is not a valid assumption by itself, (3) is a very good approximation of 3.2(7), particularly if r_{eD} is large. Similarly, if we were to compute the pressure responses at the center of the fracture ($r_D = 0$), then 3.2(7) simplifies to

$$\bar{p}_D = \frac{1}{s}\int_0^1\left[K_0\left(\sqrt{u}r'_D\right) + I_0\left(\sqrt{u}r'_D\right)\frac{K_1\left(\sqrt{u}r_{eD}\right)}{I_1\left(\sqrt{u}r_{eD}\right)}\right]dr'_D. \tag{7}$$

4.9. Flow in rectangular parallelepipeds. Some applications

We pick up on the results given in §3.3. Our main objective is to recast the solutions given there so as to render them suitable for computations.

I. *Vertical fracture, all boundaries sealed*

We begin with the pressure distribution given by

$$\bar{p}_D(x_D, y_D) = \frac{\pi}{x_{eD}s}\left[\frac{ch\sqrt{u}[y_{eD} - |y_D - y_{wD}|) + ch\sqrt{u}[y_{eD} - (y_D + y_{wD})]}{\sqrt{u}\; sh\sqrt{u}y_{eD}}\right.$$
$$+ \frac{2x_{eD}}{\pi}\sum_{k=1}^{\infty}\frac{1}{k}\sin k\pi\frac{1}{x_{eD}}\cos k\pi\frac{x_{wD}}{x_{eD}}\cos k\pi\frac{x_D}{x_{eD}} \tag{1}$$
$$\left.\frac{ch\sqrt{u+\frac{k^2\pi^2}{x_{eD}^2}}(y_{eD} - |y_D - y_{wD}|) + ch\sqrt{u+\frac{k^2\pi^2}{x_{eD}^2}}[y_{eD} - (y_D + y_{wD})]}{\sqrt{u+\frac{k^2\pi^2}{x_{eD}^2}}sh\sqrt{u+\frac{k^2\pi^2}{x_{eD}^2}}y_{eD}}\right].$$

The origin of the coordinate system for (1) is chosen to be the bottom, left-hand corner of the parallelepiped. The center of the fracture in the $x - y$ plane is given by x_{wD} and y_{wD}, and the lengths of the sides of the rectangular drainage region are x_{eD} and y_{eD}. $ch(x)$ and $sh(x)$ denote the hyperbolic cosine and hyperbolic sine functions, respectively.

A. The small-s formulation.

Because the computation of the hyperbolic terms in (1) may pose difficulties for small values of s (large times), we write

$$\frac{ch\sqrt{a}\,(y_{eD} - \tilde{y}_D)}{sh\sqrt{a}y_{eD}} = \left\{\exp\left(-\sqrt{a}\tilde{y}_D\right) + \exp\left[-\sqrt{a}\,(2y_{eD} - \tilde{y}_D)\right]\right\}$$
$$\left[1 + \sum_{m=1}^{\infty}\exp\left(-2m\sqrt{a}y_{eD}\right)\right] \tag{2}$$

If (2) is used to compute the ratios of hyperbolic functions, then the pressure distribution can be computed by (1) for times $t_{AD} \geq 10^{-2}$, where t_{AD} is the normalized time based on the drainage area and is given by

$$t_{AD} = t_D/A_D. \tag{3}$$

Here, $A_D = A/\ell^2 = x_e y_e/L_f^2$. This procedure also takes into account the possibility that $\sqrt{a}y_{eD}$ may be small.

B. The large-s formulation.

For large values of s (small times), we rewrite (1) in the following form:

$$\overline{p}_D = \overline{p}_{Di} + \overline{p}_{Db}, \tag{4}$$

where $\overline{p}_{Di}$ is the solution given by 4.6(6) for a vertically-fractured well in a slab reservoir, and $\overline{p}_{Db}$ represents the contribution of the boundaries in the x and y directions. The expression for p_{Db} is obtained as follows.

Let $\epsilon_k = \sqrt{u + k^2\pi^2/x_{eD}^2}$, $\widetilde{y}_{D1} = |y_D - y_{wD}|$, and $\widetilde{y}_{D2} = y_D + y_{wD}$. Using (2), we recast (1) in the following form:

$$\overline{p}_D = \overline{p}_{D1} + \overline{p}_{Db1} + \overline{p}_{Db2}, \tag{5}$$

where

$$\overline{p}_{D1} = \frac{2}{s}\sum_{k=1}^{\infty} \sin k\pi\frac{1}{x_{eD}} \cos k\pi\frac{x_D}{x_{eD}} \cos k\pi\frac{x_{wD}}{x_{eD}} \frac{\exp\left(-\epsilon_k \widetilde{y}_{D1}\right)}{k\epsilon_k}, \tag{6}$$

$$\begin{aligned}\overline{p}_{Db1} = {} & \frac{\pi}{x_{eD}s\sqrt{u}}\Big\{\exp\left(-\sqrt{u}\widetilde{y}_{D2}\right) + \exp\left[-\sqrt{u}\left(2y_{eD} - \widetilde{y}_{D2}\right)\right] \\ & + \exp\left(-\sqrt{u}\widetilde{y}_{D1}\right) + \exp\left[-\sqrt{u}\left(2y_{eD} - \widetilde{y}_{D1}\right)\right]\Big\} \\ & \left[1 + \sum_{m=1}^{\infty} \exp\left(-2m\sqrt{u}y_{eD}\right)\right],\end{aligned} \tag{7}$$

and

$$\begin{aligned}\overline{p}_{Db2} = {} & \frac{2}{s}\sum_{k=1}^{\infty} \frac{\sin k\pi\frac{1}{x_{eD}} \cos k\pi\frac{x_D}{x_{eD}} \cos k\pi\frac{x_{wD}}{x_{eD}}}{k\epsilon_k} \\ & \Big\{\Big\{\exp\left(-\epsilon_k\widetilde{y}_{D2}\right) + \exp\left[-\epsilon_k\left(2y_{eD} - \widetilde{y}_{D2}\right)\right] \\ & + \exp\left[-\epsilon_k\left(2y_{eD} - \widetilde{y}_{D1}\right)\right]\Big\}\left[1 + \sum_{m=1}^{\infty} \exp\left(-2m\epsilon_k y_{eD}\right)\right] \\ & + \exp\left(-\epsilon_k\widetilde{y}_{D1}\right)\sum_{m=1}^{\infty} \exp\left(-2m\epsilon_k y_{eD}\right)\Big\}.\end{aligned} \tag{8}$$

We note that $\overline{p}_{D1}$ given by (6) can be written as

$$\overline{p}_{D1} = \frac{\pi}{x_{eD} s} \sum_{k=1}^{\infty} \int_{x_{wD}-1}^{x_{wD}+1} \frac{\cos k\pi \frac{x'_{wD}}{x_{eD}} \cos k\pi \frac{x_D}{x_{eD}}}{\epsilon_k} \exp\left(-\epsilon_k \tilde{y}_{D1}\right) dx'_{wD}. \tag{9}$$

If we use the results in §4.2, we can write (9) in the following form:

$$\begin{aligned}\overline{p}_{D1} =& \frac{1}{2s} \int_{-1}^{+1} \sum_{k=-\infty}^{+\infty} \left\{ K_0 \left[\sqrt{(x_D - x_{wD} - 2kx_{eD} - \alpha)^2 + (y_D - y_{wD})^2} \sqrt{u} \right] \right. \\ & \left. + K_0 \left[\sqrt{(x_D + x_{wD} - 2kx_{eD} - \alpha)^2 + (y_D - y_{wD})^2} \sqrt{u} \right] \right\} d\alpha \\ & - \frac{\pi \exp\left(-\sqrt{u}|y_D - y_{wD}|\right)}{x_{eD} s \sqrt{u}}.\end{aligned} \tag{10}$$

Note that (10) can be written as

$$\overline{p}_{D1} = \overline{p}_{Di} + \overline{p}_{Db3}, \tag{11}$$

where $\overline{p}_{Di}$ is the vertical-fracture solution of a reservoir that is infinite in areal extent and is given by

$$\overline{p}_{Di} = \frac{1}{2s} \int_{-1}^{+1} K_0 \left[\sqrt{(x_D - x_{wD} - \alpha)^2 + (y_D - y_{wD})^2} \sqrt{u} \right] d\alpha, \tag{12}$$

and $\overline{p}_{Db3}$ is given by

$$\begin{aligned}\overline{p}_{Db3} =& \frac{1}{2s} \int_{-1}^{+1} K_0 \left[\sqrt{(x_D + x_{wD} - \alpha)^2 + (y_D - y_{wD})^2} \sqrt{u} \right] d\alpha \\ & + \frac{1}{2s} \sum_{k=1}^{\infty} \int_{-1}^{+1} \left\{ K_0 \left[\sqrt{(x_D - x_{wD} - 2kx_{eD} - \alpha)^2 + (y_D - y_{wD})^2} \sqrt{u} \right] \right. \\ & + K_0 \left[\sqrt{(x_D + x_{wD} - 2kx_{eD} - \alpha)^2 + (y_D - y_{wD})^2} \sqrt{u} \right] \\ & + K_0 \left[\sqrt{(x_D - x_{wD} + 2kx_{eD} - \alpha)^2 + (y_D - y_{wD})^2} \sqrt{u} \right] \\ & \left. + K_0 \left[\sqrt{(x_D + x_{wD} + 2kx_{eD} - \alpha)^2 + (y_D - y_{wD})^2} \sqrt{u} \right] \right\} d\alpha \\ & - \frac{\pi \exp\left(-\sqrt{u}|y_D - y_{wD}|\right)}{x_{eD} s \sqrt{u}}.\end{aligned} \tag{13}$$

Therefore, the pressure distribution resulting from production via a vertically-fractured well in a reservoir that is a closed parallelepiped is given by

$$\overline{p}_D = \overline{p}_{Di} + \overline{p}_{Db}, \tag{14}$$

where $\overline{p}_{Di}$ is given by (12) and $\overline{p}_{Db}$ is given by

$$\overline{p}_{Db} = \overline{p}_{Db1} + \overline{p}_{Db2} + \overline{p}_{Db3}. \tag{15}$$

In (15), $\overline{p}_{Db1}, \overline{p}_{Db2}$, and $\overline{p}_{Db3}$ are given, respectively, by (7), (8), and (13).

The integrals appearing in (12) and (13) can be computed along the lines suggested in §4.1. For example, for $|x_D - x_{wD}| \le 1$, and $y_D = y_{wD}$ (along the fracture plane), (13) can be written as

$$\begin{aligned}\overline{p}_{Db3} = \frac{1}{2s\sqrt{u}} \Bigg\{ &\int_0^{\sqrt{u}(x_D + x_{wD} + 1)} K_0(\xi) d\xi - \int_0^{\sqrt{u}(x_D + x_{wD} - 1)} K_0(\xi) d\xi \\ &+ \sum_{k=1}^{\infty} \Bigg[\int_0^{\sqrt{u}(2kx_{eD} - x_D + x_{wD} + 1)} K_0(\xi) d\xi - \int_0^{\sqrt{u}(2kx_{eD} - x_D + x_{wD} - 1)} K_0(\xi) d\xi \\ &+ \int_0^{\sqrt{u}(2kx_{eD} + x_D - x_{wD} + 1)} K_0(\xi) d\xi - \int_0^{\sqrt{u}(2kx_{eD} + x_D - x_{wD} - 1)} K_0(\xi) d\xi \\ &+ \int_0^{\sqrt{u}(2kx_{eD} - x_D - x_{wD} + 1)} K_0(\xi) d\xi - \int_0^{\sqrt{u}(2kx_{eD} - x_D - x_{wD} - 1)} K_0(\xi) d\xi \\ &+ \int_0^{\sqrt{u}(2kx_{eD} + x_D + x_{wD} + 1)} K_0(\xi) d\xi - \int_0^{\sqrt{u}(2kx_{eD} + x_D + x_{wD} - 1)} K_0(\xi) d\xi \Bigg] \Bigg\} \\ &- \frac{\pi}{x_{eD} s \sqrt{u}}.\end{aligned} \tag{16}$$

It is interesting that we have been able to "extract" the solution for the infinite domain from (1); see (12). The development given here is useful, from a computational viewpoint, for large values of s.

C. Large- and small-s approximations.

We have already presented a large-s approximation in 4.7(16). The same approximation can be obtained from (14) by noting that as $s \to \infty$, the leading term of (14) is $\overline{p}_{Di}$ and then using the result given in 4.1(7).

A small-s approximation is derived as follows. The pressure distribution is

$$\begin{aligned}\overline{p}_D = &H + \frac{2}{s} \sum_{k=1}^{\infty} \sin k\pi \frac{1}{x_{eD}} \cos k\pi \frac{x_{wD}}{x_{eD}} \\ &\cos k\pi \frac{x_D}{x_{eD}} \frac{ch\epsilon_k(y_{eD} - \tilde{y}_{D1}) + ch\epsilon_k(y_{eD} - \tilde{y}_{D2})}{k\epsilon_k \; sh \; \epsilon_k y_{eD}},\end{aligned} \tag{17}$$

where

$$H = \frac{\pi}{x_{eD} s} \left\{ \frac{ch\sqrt{u}(y_{eD} - \tilde{y}_{D1}) + ch\sqrt{u}(y_{eD} - \tilde{y}_{D2})}{\sqrt{u} \; sh\sqrt{u} y_{eD}} \right\}. \tag{18}$$

From Gradshteyn and Rhyzhik [29] (see 1.445–2) we have

$$\sum_{k=1}^{\infty} \frac{\cos kx}{k^2 + a^2} = \frac{\pi}{2a} \frac{ch \; a(\pi - x)}{sh \; a\pi} - \frac{1}{2a^2}; \; 0 \le x \le 2\pi, \tag{19}$$

and therefore we may write

$$\frac{1}{\sqrt{u}}\left[\frac{ch\sqrt{u}\,(y_{eD}-\widetilde{y}_D)}{sh\sqrt{u}\;y_{eD}}\right]=\frac{2}{y_{eD}}\sum_{m=1}^{\infty}\frac{\cos m\pi\frac{\widetilde{y}_D}{y_{eD}}}{\epsilon_m^2}+\frac{1}{uy_{eD}}, \tag{20}$$

where $\epsilon_m=\sqrt{u+m^2\pi^2/y_{eD}^2}$. Using (20), we can recast (18) in the following form:

$$H=\frac{2\pi}{x_{eD}y_{eD}su}+\frac{2\pi}{x_{eD}y_{eD}s}\sum_{m=1}^{\infty}\frac{\cos m\pi\frac{\widetilde{y}_{D1}}{y_{eD}}+\cos m\pi\frac{\widetilde{y}_{D2}}{y_{eD}}}{\epsilon_m^2}. \tag{21}$$

For small s, replacing u by s and $s+a$ by a, and noting the relation given in Gradshteyn and Ryzhik [30] (see 1.443–3),

$$\sum_{k=1}^{\infty}\frac{\cos kx}{k^2}=\frac{\pi^2}{6}-\frac{\pi x}{2}+\frac{x^2}{4};\quad [0\le x\le 2\pi], \tag{22}$$

we can approximate H, given by (21), as

$$\begin{aligned}\lim_{s\to 0}\left\{H\right\}&=\frac{2\pi}{x_{eD}y_{eD}s^2}+\frac{2y_{eD}}{\pi x_{eD}s}\sum_{m=1}^{\infty}\frac{1}{m^2}\left(\cos m\pi\frac{\widetilde{y}_{D1}}{y_{eD}}+\cos m\pi\frac{\widetilde{y}_{D2}}{y_{eD}}\right)\\&=\frac{2\pi}{x_{eD}y_{eD}s^2}+\frac{2\pi y_{eD}}{x_{eD}s}\left(\frac{1}{3}-\frac{\widetilde{y}_{D1}+\widetilde{y}_{D2}}{2y_{eD}}+\frac{y_{D1}^2+y_{D2}^2}{4y_{eD}^2}\right).\end{aligned} \tag{23}$$

The long-time approximation of the second term in (17) is obtained by assuming $u<<k^2\pi^2/x_{eD}^2$. Thus, application of the Inversion Theorem for the Laplace transformation to (17) yields

$$\begin{aligned}p_D&=2\pi t_{AD}+2\pi\frac{y_{eD}}{x_{eD}}\left(\frac{1}{3}-\frac{\widetilde{y}_{D1}+\widetilde{y}_{D2}}{2y_{eD}}+\frac{\widetilde{y}_{D1}^2+\widetilde{y}_{D2}^2}{4y_{eD}^2}\right)\\&+\frac{2x_{eD}}{\pi}\sum_{k=1}^{\infty}\frac{1}{k^2}\sin k\pi\frac{1}{x_{eD}}\cos k\pi\frac{x_{wD}}{x_{eD}}\\&\cos k\pi\frac{x_D}{x_{eD}}\frac{chk\pi\frac{(y_{eD}-\widetilde{y}_{D1})}{x_{eD}}+chk\pi\frac{(y_{eD}-\widetilde{y}_{D2})}{x_{eD}}}{shk\pi\frac{y_{eD}}{x_{eD}}}.\end{aligned} \tag{24}$$

II. *Horizontal well, all boundaries impermeable*

As shown in §3.3, the dimensionless pressure for a horizontal well in an isotropic reservoir ($k=k_x=k_y=k_z$) with all boundaries impermeable is given by

$$\overline{p}_D=\overline{p}_{fD}+\overline{F}_1, \tag{25}$$

where $\overline{p}_{fD}$ is the vertical-fracture solution given by (1) and $\overline{F}_1$ is given by

$$\begin{aligned}\overline{F}_1 =& \frac{2\pi}{x_{eD}s}\sum_{n=1}^{\infty}\cos n\pi z_D \cos n\pi z_{wD} \\ & \frac{ch\epsilon_n(y_{eD}-\widetilde{y}_{D1})+ch\epsilon_n(y_{eD}-\widetilde{y}_{D2})}{\epsilon_n sh\epsilon_n y_{eD}} \\ & +\frac{4}{s}\sum_{n=1}^{\infty}\cos n\pi z_D \cos n\pi z_{wD}\sum_{k=1}^{\infty}\frac{1}{k}\sin k\pi\frac{1}{x_{eD}}\cos k\pi\frac{x_{wD}}{x_{eD}}\cos k\pi\frac{x_D}{x_{eD}} \\ & \frac{ch\epsilon_{k,n}(y_{eD}-\widetilde{y}_{D1})+ch\epsilon_{k,n}(y_{eD}-\widetilde{y}_{D2})}{\epsilon_{k,n}sh\epsilon_{k,n}y_{eD}}.\end{aligned} \tag{26}$$

Here, $\epsilon_n = \sqrt{u+n^2\pi^2L_D^2}$, $\epsilon_{k,n} = \sqrt{u+k^2\pi^2/x_{eD}^2+n^2\pi^2L_D^2}$, $\widetilde{y}_{D1} = |y_D - y_{wD}|$, $\widetilde{y}_{D2} = y_D + y_{wD}$, and z_D and L_D are defined by 4.6(2) and 4.6(3), respectively. In (25), the bottom, left-hand corner of the reservoir is assumed to be the origin. The location of the center of the well is (x_{wD}, y_{wD}, z_{wD}). The sides of the rectangular region are x_{eD} and y_{eD}. $ch(x)$ and $sh(x)$ denote the hyperbolic cosine and the hyperbolic sine functions, respectively.

Points pertinent to the computation of the fracture solution, $\overline{p}_{fD}$, were discussed above. Computational issues pertinent to the computation of $\overline{F}_1$ are given below.

A. The small-s formulation for $\overline{F}_1$.

Using the procedure for the computation of the fracture solution, $\overline{p}_{fD}$, for $t_{AD} \geq 10^{-2}$, $\overline{F}_1$ can be computed from (26) without difficulty provided that the relation given in (2) is used to compute the ratio $ch(\sqrt{u}\alpha)/sh(\sqrt{u}\beta)$.

B. The large-s formulation for $\overline{F}_1$.

Using the relation given by (2), we have

$$\overline{F}_1 = \frac{2\pi}{x_{eD}s}\sum_{n=1}^{\infty}\frac{\cos n\pi z_D \cos n\pi z_{wD}}{\epsilon_n}\exp\left(-\epsilon_n\widetilde{y}_{D1}\right)+\overline{F}_2+\overline{F}_{b1}+\overline{F}_{b2}, \tag{27}$$

where

$$\begin{aligned}\overline{F}_2 =& \frac{4}{s}\sum_{n=1}^{\infty}\cos n\pi z_D \cos n\pi z_{wD}\sum_{k=1}^{\infty}\frac{\sin k\pi\frac{1}{x_{eD}}\cos k\pi\frac{x_{wD}}{x_{eD}}\cos k\pi\frac{x_D}{x_{eD}}}{k\epsilon_{k,n}} \\ & \exp\left(-\epsilon_{k,n}\widetilde{y}_{D1}\right),\end{aligned} \tag{28}$$

$$\begin{aligned}\overline{F}_{b1} =& \frac{2\pi}{x_{eD}s}\sum_{n=1}^{\infty}\frac{\cos n\pi z_D \cos n\pi z_{wD}}{\epsilon_n}\left\{\left[e^{-\epsilon_n\widetilde{y}_{D2}}+e^{-\epsilon_n(2y_{eD}-\widetilde{y}_{D2})}\right.\right. \\ & \left.+e^{-\epsilon_n(2y_{eD}-\widetilde{y}_{D1})}\right]\left[1+\sum_{m=1}^{\infty}\exp\left(-2m\epsilon_n y_{eD}\right)\right] \\ & \left.+e^{-\epsilon_n\widetilde{y}_{D1}}\sum_{m=1}^{\infty}\exp\left(-2m\epsilon_n y_{eD}\right)\right\},\end{aligned} \tag{29}$$

and

$$\overline{F}_{b2} = \frac{4}{s}\sum_{n=1}^{\infty}\cos n\pi z_D \cos n\pi z_{wD}\sum_{k=1}^{\infty}\frac{1}{k}\frac{\sin k\pi\frac{1}{x_{eD}}\cos k\pi\frac{x_D}{x_{eD}}\cos k\pi\frac{x_{wD}}{x_{eD}}}{\epsilon_{k,n}}$$
$$\left\{\left[e^{-\epsilon_{k,n}\widetilde{y}_{D2}} + e^{-\epsilon_{k,n}\left(2y_{eD}-\widetilde{y}_{D2}\right)} + e^{-\epsilon_{k,n}\left(2y_{eD}-\widetilde{y}_{D1}\right)}\right]\right. \tag{30}$$
$$\left.\left[1+\sum_{m=1}^{\infty}\exp\left(-2m\epsilon_{k,n}y_{eD}\right)\right] + e^{-\epsilon_{k,n}\widetilde{y}_{D1}}\sum_{m=1}^{\infty}\exp\left(-2m\epsilon_{k,n}y_{eD}\right)\right\}.$$

Let us now consider $\overline{F}_2$ given by (28). We may write

$$\overline{F}_2 = \frac{4}{s}\sum_{n=1}^{\infty}\cos n\pi z_D \cos n\pi z_{wD}\frac{\pi}{2x_{eD}}\int_{x_{wD}-1}^{x_{wD}+1}\sum_{k=1}^{\infty}\frac{\cos k\pi\frac{x'_{wD}}{x_{eD}}\cos k\pi\frac{x_D}{x_{eD}}}{\epsilon_{k,n}} \tag{31}$$
$$\exp\left(-\epsilon_{k,n}\widetilde{y}_{D1}\right)dx'_{wD}.$$

Using the relation given in §4.2, we can put $\overline{F}_2$ in the following form:

$$\overline{F}_2 = \frac{1}{s}\sum_{n=1}^{\infty}\cos n\pi z_D \cos n\pi z_{wD}$$
$$\int_{-1}^{+1}\sum_{k=-\infty}^{+\infty}\left\{K_0\left[\sqrt{(x_D - x_{wD} - 2kx_{eD} - \alpha)^2 + (y_D - y_{wD})^2\epsilon_n}\right]\right. \tag{32}$$
$$\left.+K_0\left[\sqrt{(x_D + x_{wD} - 2kx_{eD} - \alpha)^2 + (y_D - y_{wD})^2\epsilon_n}\right]\right\}d\alpha$$
$$-\frac{2\pi}{x_{eD}s}\sum_{n=1}^{\infty}\frac{\cos n\pi z_D \cos n\pi z_{wD}}{\epsilon_n}\exp\left(-\epsilon_n|y_D - y_{wD}|\right).$$

We can then write

$$\overline{F}_2 = \overline{F} + \overline{F}_{b3} - \frac{2\pi}{x_{eD}s}\sum_{n=1}^{\infty}\frac{\cos n\pi z_D \cos n\pi z_{wD}}{\epsilon_n}\exp\left(-\epsilon_n|y_D - y_{wD}|\right), \tag{33}$$

where

$$\overline{F} = \frac{1}{s}\sum_{n=1}^{\infty}\cos n\pi z_D \cos n\pi z_{wD} \tag{34}$$
$$\int_{-1}^{+1}K_0\left[\sqrt{(x_D - x_{wD} - \alpha)^2 + (y_D - y_{wD})^2\epsilon_n}\right]d\alpha,$$

and

$$
\begin{aligned}
\overline{F}_{b3} = & \frac{1}{s}\sum_{n=1}^{\infty} \cos n\pi z_D \cos n\pi z_{wD} \\
& \left\{ \int_{-1}^{+1} K_0\left[\sqrt{(x_D + x_{wD} - \alpha)^2 + (y_D - y_{wD})^2}\epsilon_n\right] d\alpha \right. \\
& + \sum_{k=1}^{\infty}\int_{-1}^{+1}\left\{ K_0\left[\sqrt{(x_D - x_{wD} - 2kx_{eD} - \alpha)^2 + (y_D - y_{wD})^2}\epsilon_n\right] \right. \\
& + K_0\left[\sqrt{(x_D + x_{wD} - 2kx_{eD} - \alpha)^2 + (y_D - y_{wD})^2}\epsilon_n\right] \\
& + K_0\left[\sqrt{(x_D - x_{wD} + 2kx_{eD} - \alpha)^2 + (y_D - y_{wD})^2}\epsilon_n\right] \\
& \left.\left. + K_0\left[\sqrt{(x_D + x_{wD} + 2kx_{eD} - \alpha)^2 + (y_D - y_{wD})^2}\epsilon_n\right] \right\} d\alpha \right\}.
\end{aligned}
\tag{35}
$$

From (27) and (33), we obtain

$$\overline{F}_1 = \overline{F} + \overline{F}_{b1} + \overline{F}_{b2} + \overline{F}_{b3}. \tag{36}$$

Therefore, the horizontal well solution, $\overline{p}_D$, given by (25) can be written as

$$\overline{p}_D = \overline{p}_{fD} + \overline{F} + \overline{F}_b, \tag{37}$$

where

$$\overline{F}_b = \overline{F}_{b1} + \overline{F}_{b2} + \overline{F}_{b3}. \tag{38}$$

The integrals in (34) and (35) may be computed by the relations given in §4.1. For example, for $|x_D - x_{wD}| \leq 1$ and $y_D = y_{wD}$, (34) and (35) can be written, respectively, as

$$
\begin{aligned}
\overline{F}(x_D, x_{wD}, z_D, z_{wD}, L_D) = & \frac{1}{s}\sum_{n=1}^{\infty} \frac{\cos n\pi z_D \cos n\pi z_{wD}}{\epsilon_n} \\
& \left[\int_0^{\epsilon_n[1-(x_D - x_{wD})]} K_0(\xi)d\xi + \int_0^{\epsilon_n[1+(x_D - x_{wD})]} K_0(\xi)d\xi \right],
\end{aligned}
\tag{39}
$$

and

$$\overline{F}_{b3} = \frac{1}{s}\sum_{n=1}^{\infty}\frac{\cos n\pi z_D \cos n\pi z_{wD}}{\epsilon_n}$$
$$\left\{\int_0^{\epsilon_n(x_D+x_{wD}+1)} K_0(\xi)d\xi - \int_0^{\epsilon_n(x_D+x_{wD}-1)} K_0(\xi)d\xi\right.$$
$$+\sum_{k=1}^{\infty}\left[\int_0^{\epsilon_n(2kx_{eD}-x_D+x_{wD}+1)} K_0(\xi)d\xi - \int_0^{\epsilon_n(2kx_{eD}-x_D+x_{wD}-1)} K_0(\xi)d\xi\right. \tag{40}$$
$$+\int_0^{\epsilon_n(2kx_{eD}+x_D-x_{wD}+1)} K_0(\xi)d\xi - \int_0^{\epsilon_n(2kx_{eD}+x_D-x_{wD}-1)} K_0(\xi)d\xi$$
$$+\int_0^{\epsilon_n(2kx_{eD}-x_D-x_{wD}+1)} K_0(\xi)d\xi - \int_0^{\epsilon_n(2kx_{eD}-x_D-x_{wD}-1)} K_0(\xi)d\xi$$
$$\left.\left.+\int_0^{\epsilon_n(2kx_{eD}+x_D+x_{wD}+1)} K_0(\xi)d\xi - \int_0^{\epsilon_n(2kx_{eD}+x_D+x_{wD}-1)} K_0(\xi)d\xi\right]\right\}.$$

In light of the remarks in §4.5 II, we finally obtain the following alternate forms for the functions $\overline{F}$ and $\overline{F}_{b3}$ for large values of s when $|x_D - x_{wD}| \le 1$ and $y_D = y_{wD}$:

$$\overline{F} = \frac{1}{2L_D s}\sum_{n=-\infty}^{+\infty}\left[K_0\left(\frac{|z_D - z_{wD} - 2n|\sqrt{u}}{L_D}\right) + K_0\left(\frac{|z_D + z_{wD} - 2n|\sqrt{u}}{L_D}\right)\right]$$
$$-\frac{\pi}{2s\sqrt{u}} - \frac{1}{s}\sum_{n=1}^{\infty}\frac{\cos n\pi z_D \cos n\pi z_{wD}}{\epsilon_n} \tag{41}$$
$$\left\{Ki_1\left\{\epsilon_n\left[1-(x_D - x_{wD})\right]\right\} + Ki_1\left\{\epsilon_n\left[1+(x_D - x_{wD})\right]\right\}\right\},$$

and

$$
\begin{aligned}
\overline{F}_{b3} = \frac{1}{s}\sum_{n=1}^{\infty} & \frac{\cos n\pi z_D \cos n\pi z_{wD}}{\epsilon_n} \\
& \Big\{ Ki_1\left[\epsilon_n(x_D + x_{wD} + 1)\right] - Ki_1\left[\epsilon_n(x_D + x_{wD} - 1)\right] \\
& + \sum_{k=1}^{\infty} \{ Ki_1\left[\epsilon_n(2kx_{eD} - x_D + x_{wD} + 1)\right] \\
& - Ki_1\left[\epsilon_n(2kx_{eD} - x_D + x_{wD} - 1)\right] \\
& + Ki_1\left[\epsilon_n(2kx_{eD} + x_D - x_{wD} + 1)\right] \\
& - Ki_1\left[\epsilon_n(2kx_{eD} + x_D - x_{wD} - 1)\right] \\
& + Ki_1\left[\epsilon_n(2kx_{eD} - x_D - x_{wD} + 1)\right] \\
& - Ki_1\left[\epsilon_n(2kx_{eD} - x_D - x_{wD} - 1)\right] \\
& + Ki_1\left[\epsilon_n(2kx_{eD} + x_D + x_{wD} + 1)\right] \\
& - Ki_1\left[\epsilon_n(2kx_{eD} + x_D + x_{wD} - 1)\right]\} \Big\}.
\end{aligned} \tag{42}
$$

In (41) and (42), $Ki_1(x)$ is defined in 4.1(22).

C. The large- and small-s approximations.

To obtain a large-s approximation, let $u = \omega s$ and note that in (37) as $s \to \infty$, $\overline{F} >> \overline{F}_b$. Then, substituting the large-s approximations for $\overline{p}_{fD}$ given in 4.7(15) and for $\overline{F}$ given in 4.5(20), we obtain the following large-s approximation for $\overline{p}_D$:

$$
\overline{p}_D = \frac{\beta}{4L_D s} K_0 \left[\sqrt{(z_D - z_{wD})^2 / L_D^2 + (y_D - y_{wD})^2}\sqrt{\omega s} \right], \tag{43}
$$

where

$$
\beta = \begin{cases} 2 & \text{for } |x_D - x_{wD}| < 1, \\ 1 & \text{for } |x_D - x_{wD}| = 1, \\ 0 & \text{for } |x_D - x_{wD}| > 1. \end{cases} \tag{44}
$$

Application of the Inversion Theorem for the Laplace transformation to (43) yields

$$
p_D = -\frac{\beta}{8L_D} Ei \left[-\frac{(z_D - z_{wD})^2 / L_D^2 + (y_D - y_{wD})^2}{4t_D/\omega} \right]. \tag{45}
$$

To obtain a small-s approximation we first consider $\overline{F}_1$. A small-s approximation for $\overline{F}_1$ can be obtained simply by setting u to 0. Let $\dot{\epsilon}_n$ and $\dot{\epsilon}_{k,n}$ be the values of ϵ_n and $\epsilon_{k,n}$, respectively, when $u = 0$. Thus, $\overline{F}_1$ is given by

$$
\begin{aligned}
\overline{F}_1 = & \frac{2}{x_{eD} L_D s} \sum_{n=1}^{\infty} \frac{1}{n} \cos n\pi z_D \cos n\pi z_{wD} \frac{ch\dot{\epsilon}_n(y_{eD} - \tilde{y}_{D1}) + ch\dot{\epsilon}_n(y_{eD} - \tilde{y}_{D2})}{sh\dot{\epsilon}_n y_{eD}} \\
& + \frac{4}{s}\sum_{n=1}^{\infty} \cos n\pi z_D \cos n\pi z_{wD} \sum_{k=1}^{\infty} \frac{\sin k\pi \frac{1}{x_{eD}} \cos k\pi \frac{x_{wD}}{x_{eD}} \cos k\pi \frac{x_D}{x_{eD}}}{k \dot{\epsilon}_{k,n}} \\
& \frac{ch\dot{\epsilon}_{k,n}(y_{eD} - \tilde{y}_{D1}) + ch\dot{\epsilon}_{k,n}(y_{eD} - \tilde{y}_{D2})}{sh\dot{\epsilon}_{k,n} y_{eD}},
\end{aligned} \tag{46}
$$

and by the application of the Inversion Theorem for the Laplace transformation we obtain

$$p_D = p_{fD} + F_1, \tag{47}$$

where p_{fD} is given in (24) and F_1 is given by

$$F_1 = \frac{2}{x_{eD}L_D}\sum_{n=1}^{\infty}\frac{1}{n}\cos n\pi z_D \cos n\pi z_{wD}\frac{ch\dot{\epsilon}_n(y_{eD}-\widetilde{y}_{D1}) + ch\dot{\epsilon}_n(y_{eD}-\widetilde{y}_{D2})}{sh\dot{\epsilon}_n y_{eD}}$$
$$+4\sum_{n=1}^{\infty}\cos n\pi z_D \cos n\pi z_{wD}\sum_{k=1}^{\infty}\frac{\sin k\pi\frac{1}{x_{eD}}\cos k\pi\frac{x_{wD}}{x_{eD}}\cos k\pi\frac{x_D}{x_{eD}}}{k\dot{\epsilon}_{k,n}} \tag{48}$$
$$\frac{ch\dot{\epsilon}_{k,n}(y_{eD}-\widetilde{y}_{D1}) + ch\dot{\epsilon}_{k,n}(y_{eD}-\widetilde{y}_{D2})}{sh\dot{\epsilon}_{k,n}y_{eD}}.$$

For computational purposes it is better to replace the right-hand side of (48) by

$$F_1 = F + F_{b1} + F_{b2} + F_{b3}. \tag{49}$$

Expressions for F, F_{b1}, F_{b2}, and F_{b3} are given, respectively, by

$$F = \sum_{n=1}^{\infty}\cos n\pi z_D \cos n\pi z_{wD}$$
$$\int_{-1}^{+1}K_0\left[n\pi L_D\sqrt{(x_D - x_{wD} - \alpha)^2 + (y_D - y_{wD})^2}\right]d\alpha, \tag{50}$$

$$F_{b1} = \frac{2}{x_{eD}L_D}\sum_{n=1}^{\infty}\frac{1}{n}\cos n\pi z_D \cos n\pi z_{wD}\left\{\left[e^{-n\pi L_D\widetilde{y}_{D2}}\right.\right.$$
$$\left. + e^{-n\pi L_D(2y_{eD}-\widetilde{y}_{D1})} + e^{-n\pi L_D(2y_{eD}-\widetilde{y}_{D2})}\right]$$
$$\left[1 + \sum_{m=1}^{\infty}\exp(-2mn\pi L_D y_{eD})\right] \tag{51}$$
$$\left. + e^{-n\pi L_D\widetilde{y}_{D1}}\sum_{m=1}^{\infty}\exp(-2mn\pi L_D y_{eD})\right\},$$

$$F_{b2} = 4\sum_{n=1}^{\infty}\cos n\pi z_D \cos n\pi z_{wD}\sum_{k=1}^{\infty}\frac{1}{k}\frac{\sin k\pi\frac{1}{x_{eD}}\cos k\pi\frac{x_D}{x_{eD}}\cos k\pi\frac{x_{wD}}{x_{eD}}}{\dot{\epsilon}_{k,n}}$$
$$\left\{\left[e^{-\dot{\epsilon}_{k,n}\widetilde{y}_{D2}} + e^{-\dot{\epsilon}_{k,n}(2y_{eD}-\widetilde{y}_{D1})}\right.\right.$$
$$\left. + e^{-\dot{\epsilon}_{k,n}(2y_{eD}-\widetilde{y}_{D2})}\right]\left[1 + \sum_{m=1}^{\infty}\exp\left(-2m\dot{\epsilon}_{k,n}y_{eD}\right)\right] \tag{52}$$
$$\left. + e^{-\dot{\epsilon}_{k,n}\widetilde{y}_{D1}}\sum_{m=1}^{\infty}\exp\left(-2m\dot{\epsilon}_{k,n}y_{eD}\right)\right\},$$

and

$$F_{b3} = \sum_{n=1}^{\infty} \cos n\pi z_D \cos n\pi z_{wD}$$

$$\left\{ \int_{-1}^{+1} K_0 \left[n\pi L_D \sqrt{(x_D + x_{wD} - \alpha)^2 + (y_D - y_{wD})^2} \right] d\alpha \right.$$

$$+ \sum_{k=1}^{\infty} \int_{-1}^{+1} \left\{ K_0 \left[n\pi L_D \sqrt{(x_D - x_{wD} - 2kx_{eD} - \alpha)^2 + (y_D - y_{wD})^2} \right] \right. \tag{53}$$

$$+ K_0 \left[n\pi L_D \sqrt{(x_D + x_{wD} - 2kx_{eD} - \alpha)^2 + (y_D - y_{wD})^2} \right]$$

$$+ K_0 \left[n\pi L_D \sqrt{x_D - x_{wD} + 2kx_{eD} - \alpha)^2 + (y_D - y_{wD})^2} \right]$$

$$\left. \left. + K_0 \left[n\pi L_D \sqrt{(x_D + x_{wD} + 2kx_{eD} - \alpha)^2 + (y_D - y_{wD})^2} \right] \right\} d\alpha \right\}.$$

The computation of the functions given above is straightforward. To compute the functions F and F_{b3} given by (50) and (53), respectively, the relations given in §4.1 are useful.

III. *Inclined well, one boundary at the initial pressure*

As noted in §3.3 II, the solution for the pressure distribution is given by

$$\overline{p}_D = \frac{2\pi}{x_{eD} h_{fD} s} \int_{-h_{fD}/2}^{+h_{fD}/2} \sum_{k=1}^{\infty} \cos(2k-1)\frac{\pi}{2}\frac{x_D}{x_{eD}} \cos(2k-1)\frac{\pi}{2}\frac{\tilde{x}_{wD}}{x_{eD}}$$

$$\left[\frac{ch\epsilon_{2k-1}(y_{eD} - \tilde{y}_{D1}) + ch\epsilon_{2k-1}(y_{eD} - \tilde{y}_{D2})}{\epsilon_{2k-1} sh\left(\epsilon_{2k-1} y_{eD}\right)} \right.$$

$$\left. + 2\sum_{n=1}^{\infty} \cos n\pi \frac{z_D}{z_{eD}} \cos n\pi \frac{\tilde{z}_{wD}}{z_{eD}} \frac{ch\epsilon_{2k-1,n}(y_{eD} - \tilde{y}_{D1}) + ch\epsilon_{2k-1,n}(y_{eD} - \tilde{y}_{D2})}{\epsilon_{2k-1,n} sh\epsilon_{2k-1,n} y_{eD}} \right] d\alpha. \tag{54}$$

Here,

$$\epsilon_{2k-1} = \sqrt{u + (2k-1)^2\pi^2/(4x_{eD}^2)},$$

$$\epsilon_{2k-1,n} = \sqrt{u + (2k-1)^2\pi^2/(4x_{eD}^2) + n^2\pi^2/z_{eD}^2},$$

$\tilde{x}_{wD} = x_{wD} + \alpha \sin\psi$, $\tilde{z}_{wD} = z_{wD} + \alpha\cos\psi$, $\tilde{y}_{D1} = |y_D - y_{wD}|$, and $\tilde{y}_{D2} = y_D + y_{wD}$. We now note a few computational issues.

A. A small-s formulation.

For small s ($t_{AD} \geq 10^{-2}$, where t_{AD} is given in (3)), the pressure distribution can be computed from (54) by using (2).

B. A large-s formulation.

For large s, noting the relation in (2), we may write (54) as

$$\overline{p}_D = \overline{p}_{D1} + \overline{p}_{D3} + \overline{p}_{Db1} + \overline{p}_{Db2}. \tag{55}$$

The various expressions for the right-hand side of (55) are given below:

$$\overline{p}_{D1} = \frac{2\pi}{x_{eD} h_{fD} s} \int_{-h_{fD}/2}^{+h_{fD}/2} \sum_{k=1}^{\infty} \cos(2k-1)\frac{\pi}{2}\frac{x_D}{x_{eD}} \cos(2k-1)\frac{\pi}{2}\frac{\widetilde{x}_{wD}}{x_{eD}} \frac{\exp(-\epsilon_{2k-1}\widetilde{y}_{D1})}{\epsilon_{2k-1}} d\alpha. \tag{56}$$

$$\overline{p}_{D3} = \frac{4\pi}{x_{eD} h_{fD} s} \int_{-h_{fD}/2}^{+h_{fD}/2} \sum_{k=1}^{\infty} \cos(2k-1)\frac{\pi}{2}\frac{x_D}{x_{eD}} \cos(2k-1)\frac{\pi}{2}\frac{\widetilde{x}_{wD}}{x_{eD}} \sum_{n=1}^{\infty} \cos n\pi\frac{z_D}{z_{eD}} \cos n\pi\frac{\widetilde{z}_{wD}}{z_{eD}} \exp(-\epsilon_{2k-1,n}\widetilde{y}_{D1}) \frac{d\alpha}{\epsilon_{2k-1,n}}. \tag{57}$$

$$\overline{p}_{Db1} = \frac{2\pi}{x_{eD} h_{fD} s} \int_{-h_{fD}/2}^{+h_{fD}/2} \sum_{k=1}^{\infty} \cos(2k-1)\frac{\pi}{2}\frac{x_D}{x_{eD}} \cos(2k-1)\frac{\pi}{2}\frac{\widetilde{x}_{wD}}{x_{eD}} f(\epsilon_{2k-1}) \frac{d\alpha}{\epsilon_{2k-1}}. \tag{58}$$

$$\overline{p}_{Db2} = \frac{4\pi}{x_{eD} h_{fD} s} \int_{-h_{fD}/2}^{+h_{fD}/2} \sum_{k=1}^{\infty} \cos(2k-1)\frac{\pi}{2}\frac{x_D}{x_{eD}} \cos(2k-1)\frac{\pi}{2}\frac{\widetilde{x}_{wD}}{x_{eD}} \sum_{n=1}^{\infty} \cos n\pi\frac{z_D}{z_{eD}} \cos n\pi\frac{\widetilde{z}_{wD}}{z_{eD}} f(\epsilon_{2k-1,n}) \frac{d\alpha}{\epsilon_{2k-1,n}}. \tag{59}$$

In (58) and (59), $f(\widetilde{\epsilon})$ is given by

$$f(\widetilde{\epsilon}) = \Big\{ \exp\left[-\widetilde{\epsilon}(2y_{eD} - \widetilde{y}_{D1})\right] + \exp(-\widetilde{\epsilon}\,\widetilde{y}_{D2}) + \exp\left[-\widetilde{\epsilon}(2y_{eD} - \widetilde{y}_{D2})\right] \Big\} \left[1 + \sum_{\ell=1}^{\infty} \exp(-2\ell\widetilde{\epsilon} y_{eD})\right] + \exp(-\widetilde{\epsilon}\,\widetilde{y}_{D1}) \sum_{\ell=1}^{\infty} \exp(-2\ell\widetilde{\epsilon} y_{eD}). \tag{60}$$

Using the result in §4.4, we can write

$$\overline{p}_{D1} = \frac{1}{h_{fD} s} \int_{-h_{fD}/2}^{+h_{fD}/2} \sum_{k=-\infty}^{+\infty} (-1)^k \Big\{ K_0\left[\sqrt{(x_D - \widetilde{x}_{wD} - 2kx_{eD})^2 + (y_D - y_{wD})^2}\sqrt{u}\right] + K_0\left[\sqrt{(x_D + \widetilde{x}_{wD} - 2kx_{eD})^2 + (y_D - y_{wD})^2}\sqrt{u}\right] \Big\} d\alpha. \tag{61}$$

(61) may be rewritten as

$$\overline{p}_{D1} = \overline{p}_{Di1} + \overline{p}_{Db3}, \tag{62}$$

where

$$\overline{p}_{Di1} = \frac{1}{h_{fD}s}\int_{-h_{fD}/2}^{+h_{fD}/2} K_0\left[\sqrt{(x_D-\widetilde{x}_{wD})^2+(y_D-y_{wD})^2}\sqrt{u}\right]d\alpha, \tag{63}$$

and

$$\begin{aligned}\overline{p}_{Db3} = \frac{1}{h_{fD}s}\int_{-h_{fD}/2}^{+h_{fD}/2} &\left\{K_0\left[\sqrt{(x_D+\widetilde{x}_{wD})^2+(y_D-y_{wD})^2}\sqrt{u}\right]\right.\\ &\left.+\sum_{k=1}^{\infty}(-1)^k f_k\left[K_0(x_D,y_D,\sqrt{u})\right]\right\}d\alpha,\end{aligned} \tag{64}$$

where

$$\begin{aligned}f_k\left[K_0(x_D,y_D,\widetilde{\epsilon})\right] = {} & K_0\left[\sqrt{(x_D-\widetilde{x}_{wD}-2kx_{eD})^2+(y_D-y_{wD})^2\widetilde{\epsilon}}\right]\\ &+K_0\left[\sqrt{(x_D+\widetilde{x}_{wD}-2kx_{eD})^2+(y_D-y_{wD})^2\widetilde{\epsilon}}\right]\\ &+K_0\left[\sqrt{(x_D-\widetilde{x}_{wD}+2kx_{eD})^2+(y_D-y_{wD})^2\widetilde{\epsilon}}\right]\\ &+K_0\left[\sqrt{(x_D+\widetilde{x}_{wD}+2kx_{eD})^2+(y_D-y_{wD})^2\widetilde{\epsilon}}\right].\end{aligned} \tag{65}$$

Similarly, we can write (57) as

$$\overline{p}_{D3} = \overline{p}_{Di2} + \overline{p}_{Db4}, \tag{66}$$

where

$$\begin{aligned}\overline{p}_{Di2} = {} & \frac{2}{h_{fD}s}\int_{-h_{fD}/2}^{+h_{fD}/2}\sum_{n=1}^{\infty} K_0\left[\sqrt{(x-\widetilde{x}_{wD})^2+(y_D-y_{wD})^2}\epsilon_n\right]\\ &\cos n\pi\frac{z_D}{z_{eD}}\cos n\pi\frac{\widetilde{z}_{wD}}{z_{eD}}d\alpha,\end{aligned} \tag{67}$$

and

$$\begin{aligned}\overline{p}_{Db4} = {} & \frac{2}{h_{fD}s}\int_{-h_{fD}/2}^{+h_{fD}/2}\sum_{n=1}^{\infty}\cos n\pi\frac{z_D}{z_{eD}}\cos n\pi\frac{\widetilde{z}_{wD}}{z_{eD}}\\ &\left\{K_0\left[\sqrt{(x_D+\widetilde{x}_{wD})^2+(y_D-\widetilde{y}_{wD})^2}\epsilon_n\right]\right.\\ &\left.+\sum_{k=1}^{\infty}(-1)^k f_k\left[K_0(x_D,y_D,\epsilon_n)\right]\right\}d\alpha.\end{aligned} \tag{68}$$

In (67) and (68) $f_k[K_0(x_D, y_D, \widetilde{\epsilon})]$ is given by (65) and $\epsilon_n = \sqrt{u + n^2\pi^2/z_{eD}^2}$. Thus, in essence, we can write (54) as

$$\overline{p}_D = \overline{p}_{Di} + \overline{p}_{Db}, \tag{69}$$

where

$$\overline{p}_{Di} = \overline{p}_{Di1} + \overline{p}_{Di2} \tag{70}$$

and

$$\overline{p}_{Db} = \overline{p}_{Db1} + \overline{p}_{Db2} + \overline{p}_{Db3} + \overline{p}_{Db4}. \tag{71}$$

The solution for a vertical well as well as that for a horizontal well in a reservoir that is infinite in areal extent are contained in the above solution.

IV. *Vertical well, two boundaries at the initial pressure*

Let the constant pressure boundaries be at $x_D = x_{eD}$ and $y_D = y_{eD}$. The pressure distribution is given by (see §3.3 III)

$$\begin{aligned}\overline{p}_D = \frac{2\pi}{x_{eD}s} \sum_{k=1}^{\infty} \cos(2k-1)\frac{\pi}{2}\frac{x_D}{x_{eD}} \cos(2k-1)\frac{\pi}{2}\frac{x_{wD}}{x_{eD}} \\ \frac{sh\epsilon_{2k-1}(y_{eD} - \tilde{y}_{D1}) + sh\epsilon_{2k-1}(y_{eD} - \tilde{y}_{D2})}{\epsilon_{2k-1}ch(\epsilon_{2k-1}y_{eD}},\end{aligned} \tag{72}$$

where $\epsilon_{2k-1} = \sqrt{u + (2k-1)^2\pi^2/(4x_{eD}^2)}$.

A. The small-s formulation.

Similar to the solutions discussed in I, computation of the hyperbolic terms in (72) may pose difficulties. If we use the relation

$$\begin{aligned}\frac{sh\sqrt{a}(y_{eD} - \tilde{y}_D)}{ch\sqrt{a}y_{eD}} = \left\{\exp\left(-\sqrt{a}\tilde{y}_D\right) - \exp\left[-\sqrt{a}(2y_{eD} - \tilde{y}_D)\right]\right\} \\ \left[1 + \sum_{m=1}^{\infty}(-1)^m \exp(-2m\sqrt{a}y_{eD})\right],\end{aligned} \tag{73}$$

then $\overline{p}_D$ can be computed from (72) for small values of s ($t_{AD} \geq 10^{-2}$, where t_{AD} is given in (3)).

B. The large-s formulation.

For large values of s, using (73), we may write

$$\overline{p}_D = \overline{p}_{D1} + \overline{p}_{D2}, \tag{74}$$

where

$$\overline{p}_{D1} = \frac{2\pi}{x_{eD}s} \sum_{k=1}^{\infty} \frac{\cos(2k-1)\frac{\pi}{2}\frac{x_D}{x_{eD}} \cos(2k-1)\frac{x_{wD}}{x_{eD}}}{\epsilon_{2k-1}} \exp\left(-\epsilon_{2k-1}\tilde{y}_{D1}\right) \tag{75}$$

and

$$\overline{p}_{D2} = \frac{2\pi}{x_{eD}s} \sum_{k=1}^{\infty} \frac{\cos(2k-1)\frac{\pi}{2}\frac{x_D}{x_{eD}} \cos(2k-1)\frac{x_{wD}}{x_{eD}}}{\epsilon_{2k-1}}$$

$$\Bigg\{ \Bigg\{ \exp\left(-\epsilon_{2k-1}\widetilde{y}_{D2}\right) - \exp\left[-\epsilon_{2k-1}(2y_{eD} - \widetilde{y}_{D1})\right]$$

$$- \exp\left[-\epsilon_{2k-1}\left(2y_{eD} - \widetilde{y}_{D2}\right)\right]\Bigg\} \left[1 + \sum_{m=1}^{\infty}(-1)^m \exp(-2m\epsilon_{2k-1}y_{eD})\right] \tag{76}$$

$$+ \exp(-\epsilon_{2k-1}\widetilde{y}_{D1}) \sum_{m=1}^{\infty}(-1)^m \exp(-2m\epsilon_{2k-1}y_{eD})\Bigg\}.$$

Using the result in §4.4, $\overline{p}_{D1}$ can be written as

$$\overline{p}_{D1} = \overline{p}_{Di} + \overline{p}_{D3}, \tag{77}$$

where

$$\overline{p}_{Di} = \frac{1}{s} K_0 \left[\sqrt{(x_D - x_{wD})^2 + (y_D - y_{wD})^2}\sqrt{u}\right], \tag{78}$$

and

$$\overline{p}_{D3} = \frac{1}{s}\Bigg\{ K_0 \left[\sqrt{(x_D + x_{wD})^2 + (y_D - y_{wD})^2}\sqrt{u}\right] + \sum_{n=1}^{\infty}(-1)^n f_n \left[K_0(x_D, y_D, \sqrt{u})\right]\Bigg\}, \tag{79}$$

where

$$\begin{aligned} f_n\left[K_0\left(x_D, y_D, \widetilde{\epsilon}\right)\right] = {} & K_0 \left[\sqrt{(x_D - x_{wD} - 2nx_{eD})^2 + (y_D - y_{wD})^2}\widetilde{\epsilon}\right] \\ & + K_0 \left[\sqrt{(x_D + x_{wD} - 2nx_{eD})^2 + (y_D - y_{wD})^2}\widetilde{\epsilon}\right] \\ & + K_0 \left[\sqrt{(x_D - x_{wD} + 2nx_{eD})^2 + (y_D - y_{wD})^2}\widetilde{\epsilon}\right] \\ & + K_0 \left[\sqrt{(x_D + x_{wD} + 2nx_{eD})^2 + (y_D - y_{wD})^2}\widetilde{\epsilon}\right]. \end{aligned} \tag{80}$$

Thus, again, we have

$$\overline{p}_D = \overline{p}_{Di} + \overline{p}_{Db}, \tag{81}$$

where $\overline{p}_{Db} = \overline{p}_{D2} + \overline{p}_{D3}$.

V. *Vertically-fractured well, four boundaries at the initial pressure*

Let the boundaries $x_D = 0$, $x_D = x_{eD}$, $y_D = 0$ and $y_D = y_{eD}$ be at a constant pressure. The pressure distribution can be written as (see §3.3 IV)

$$\overline{p}_D = \frac{2}{s}\sum_{k=1}^{\infty} \frac{1}{k} \sin k\pi \frac{x_D}{x_{eD}} \sin k\pi \frac{x_{wD}}{x_{eD}} \sin k\pi \frac{1}{x_{eD}} \frac{ch\epsilon_k(y_{eD} - \widetilde{y}_{D1}) - ch\epsilon_k(y_{eD} - \widetilde{y}_{D2})}{\epsilon_k sh\epsilon_k y_{eD}}, \tag{82}$$

where $\epsilon_k = \sqrt{u + k^2\pi^2/x_{eD}^2}$, $\widetilde{y}_{D1} = |y_D - y_{wD}|$, and $\widetilde{y}_{D2} = y_D + y_{wD}$.

A. The small-s formulation.

For small s, we follow the ideas noted in I.

B. The large-s formulation.

Using (2), we write

$$\overline{p}_D = \overline{p}_{D1} + \overline{p}_{D2}, \tag{83}$$

where

$$\begin{aligned}\overline{p}_{D1} &= \frac{2}{s}\sum_{k=1}^{\infty}\frac{1}{k}\sin k\pi\frac{x_D}{x_{eD}}\sin k\pi\frac{x_{wD}}{x_{eD}}\sin k\pi\frac{1}{x_{eD}}\frac{\exp(-\epsilon_k\widetilde{y}_{D1})}{\epsilon_k} \\ &= \frac{\pi}{sx_{eD}}\sum_{k=1}^{\infty}\sin k\pi\frac{x_D}{x_{eD}}\int_{x_{wD}-1}^{x_{wD}+1}\sin k\pi\frac{x'_{wD}}{x_{eD}}dx'_{wD}\frac{\exp(-\epsilon_k\widetilde{y}_{D1})}{\epsilon_k},\end{aligned} \tag{84}$$

and

$$\begin{aligned}\overline{p}_{D2} = \frac{2}{s}\sum_{k=1}^{\infty}&\frac{\sin k\pi\frac{x_D}{x_{eD}}\sin k\pi\frac{x_{wD}}{x_{eD}}\sin k\pi\frac{1}{x_{eD}}}{k\epsilon_k} \\ &\Bigg\{\Bigg\{\exp\left[-\epsilon_k(2y_{eD} - \widetilde{y}_{D1})\right] - \exp\left(-\epsilon_k\widetilde{y}_{D2}\right) \\ &- \exp\left[-\epsilon_k\left(2y_{eD} - \widetilde{y}_{D2}\right)\right]\Bigg\}\left[1 + \sum_{m=1}^{\infty}\exp(-2m\epsilon_k y_{eD})\right] \\ &+ \exp\left(-\epsilon_k\widetilde{y}_{D1}\right)\sum_{m=1}^{\infty}\exp(-2m\epsilon_k y_{eD})\Bigg\}.\end{aligned} \tag{85}$$

If we use the result given in §4.3, then

$$\begin{aligned}\overline{p}_{D1} = \frac{1}{2s}\sum_{k=-\infty}^{+\infty}\int_{-1}^{+1}&\left\{K_0\left[\sqrt{(x_D - x_{wD} - 2kx_{eD} - \alpha)^2 + (y_D - y_{wD})^2}\sqrt{u}\right]\right. \\ &\left.- K_0\left[\sqrt{(x_D + x_{wD} - 2kx_{eD} - \alpha)^2 + (y_D - y_{wD})^2}\sqrt{u}\right]\right\}d\alpha.\end{aligned} \tag{86}$$

(86) may be split into the two terms

$$\overline{p}_{Di} = \frac{1}{2s}\int_{-1}^{+1}K_0\left[\sqrt{(x_D - x_{wD} - \alpha)^2 + (y_D - y_{wD})^2}\sqrt{u}\right]d\alpha, \tag{87}$$

and

$$\begin{aligned}
\overline{p}_{D3} = &-\frac{1}{2s}\int_{-1}^{+1} K_0\left[\sqrt{(x_D + x_{wD} - \alpha)^2 + (y_D - y_{wD})^2}\sqrt{u}\right] d\alpha \\
&+ \frac{1}{2s}\sum_{k=1}^{\infty}\int_{-1}^{+1}\left\{K_0\left[\sqrt{(x_D - x_{wD} - 2kx_{eD} - \alpha)^2 + (y_D - y_{wD})^2}\sqrt{u}\right]\right. \\
&- K_0\left[\sqrt{(x_D + x_{wD} - 2kx_{eD} - \alpha)^2 + (y_D - y_{wD})^2}\sqrt{u}\right] \\
&+ K_0\left[\sqrt{(x_D - x_{wD} + 2kx_{eD} - \alpha)^2 + (y_D - y_{wD})^2}\sqrt{u}\right] \\
&\left.- K_0\left[\sqrt{(x_D + x_{wD} + 2kx_{eD} - \alpha)^2 + (y_D - y_{wD})^2}\sqrt{u}\right]\right\} d\alpha.
\end{aligned} \tag{88}$$

Thus, we may write the pressure distribution in the form

$$\overline{p}_D = \overline{p}_{Di} + \overline{p}_{Db}, \tag{89}$$

where $\overline{p}_{Db} = \overline{p}_{D2} + \overline{p}_{D3}$.

References

1. Abramowitz, M. and Stegun, I. A.: *Handbook of Mathematical Functions*, Dover Publications, Inc., New York (1972), 480.
2. Luke, Y. L.: *Integrals of Bessel Functions*, McGraw Hill Book Co., Inc., New York City (1962), 60–69.
3. Abramowitz, M. and Stegun, I. A.: *op. cit.*, 480.
4. Kuchuk, F. J.: "New Methods for Estimating Parameters of Low Permeability Reservoirs," Paper *SPE/DOE* 16394 (1987).
5. Gradshteyn, I. S. and Ryzhik, I. M: *Table of Integrals, Series, and Products*, Academic Press, Inc., Orlando (1980), 683.
6. Gringarten, A. C., Ramey, H. J., Jr., and Raghavan, R.: "Unsteady-State Pressure Distributions Created by a Well With a Single Infinite-Conductivity Vertical Fracture," *Society of Petroleum Engineers Journal* (August 1974), 347–360.
7. Stakgold, I.: *Green's Functions and Boundary Value Problems*, Wiley-Interscience (1979), 462.
8. Abramowitz, M. and Stegun, I. A.: *op. cit.*, 376.
9. Uraiet, A. A., Raghavan, R., and Thomas, G. W.: "Determination of the Orientation of a Vertical Fracture by Interference Tests," *Journal of Petroleum Technology* (January 1977), 73–80.
10. Gradshteyn, I. S. and Ryzhik, I. M.: *op. cit.*, 705.
11. Gradshteyn, I. S. and Ryzhik, I. M.: *op. cit.*, 666.

12. Luke, Y. L.: *op. cit.*, 88.

13. Gradshteyn, I. S. and Ryzhik, I. M.: *op. cit.*, 41.

14. Gradshteyn, I. S. and Ryzhik, I. M.: *op. cit.*, 41.

15. Gradshteyn, I. S. and Ryzhik, I. M.: *op. cit.*, 498.

16. Gradshteyn, I. S. and Ryzhik, I. M.: *op. cit.*, 42.

17. Gradshteyn, I. S. and Ryzhik, I. M.: *op. cit.*, 419.

18. Gradshteyn, I. S. and Ryzhik, I. M.: *op. cit.*, 978.

19. Gringarten, A. C., Ramey, H. J. Jr., and Raghavan, R.: *op. cit.*, 347–360.

20. Muskat, M.: *The Flow of Homogeneous Fluids Through Porous Media,* J. W. Edwards, Inc., Ann Arbor, Michigan (1946), 263.

21. Kuchuk, F. J.: *op. cit.*. (1987).

22. Hantush, M. S.: "Aquifer Tests in Partially Penetrating Wells," *Proc.* ASCE, H. Y. 5 (1961) Vol. 37, 171.

23. Streltsova–Adams, T. D.: "Pressure Drawdown in a Well with Limited Flow Entry," *Journal of Petroleum Technology* (Nov. 1979), 1469–1476.

24. Kuchuk, F., Goode, P. A., Wilkinson, D., and Thambynayagam, R. K. M.: "Pressure-Transient Behavior of Horizontal Wells With and Without Gas Cap and an Aquifer," *SPE Formation Evaluation* (March 1991), 86–94.

25. Wilkinson, D. and Hammond, P. S.: "A Perturbation Method for Mixed Boundary Value Problems in Pressure Transient Testing," *Transport in Porous Media* (1990), **5**, 609–636.

26. Chen, H. Y.: Personal Communication (1993).

27. Gringarten, A. C. and Ramey, H. J., Jr.: "The Use of Source and Green's Functions in Solving Unsteady-Flow Problems in Reservoirs," *Society of Petroleum Engineers Journal* (Oct. 1973), 285–296.

28. Watson, G. N.: *A Treatise on the Theory of Bessel Functions*, Cambridge University Press, London (1948), 361.

29. Gradshteyn, I. S. and Ryzhik, I. M.: *op. cit.*, 40.

30. Gradshteyn, I. S. and Ryzhik, I. M.: *op. cit.*, 39.

V. Flow in Fissured and Layered Porous Media

Here, our objective is to extend the solutions discussed in Chapters III and IV to a wide class of problems. First, we begin with a discussion of variable-rate solutions using Duhamel's [1] integral formula. Second, we discuss production of fluids via a wellbore at a constant rate at the surface as opposed to a constant rate at depth. This problem is known as the "wellbore-storage problem" and was first discussed by van Everdingen and Hurst [2]. We also briefly explore the influence of a region of altered permeability around the wellbore that results from the drilling of the well. Third, we consider extending the applicability of our solutions to naturally-fractured or fissured porous media. Specifically, we examine the model considered by Barenblatt, Zheltov, and Kochina [3]. We then extend these results to more complicated visualizations of fissured porous media. Fourth, we consider flow in layered porous media wherein there is communication between the strata only via the wellbore (commingled production). Fifth, we examine methods to couple wellbore hydraulics (pipe flow) with flow in the porous medium, permitting us to explore the interaction between the porous medium and the wellbore through which the subterranean fluid must first flow to be produced at the surface.

5.1. Duhamel's [1] formula

Let p_{Du} represent the normalized response of a system that is produced at a constant rate. If this system is produced at a variable rate, then the well-known theorem of Duhamel provides the following expression for the pressure distribution:

$$\begin{aligned} p(M_D;t_D) &= p_i - \frac{\mu}{2\pi kh}\frac{\partial}{\partial t_D}\int_0^{t_D} q(t_D-\tau)p_{Du}(M_D;\tau)d\tau, \\ &= p_i - \frac{\mu}{2\pi kh}\int_0^{t_D} q(t_D-\tau)p'_{Du}(M_D;\tau)d\tau. \end{aligned} \tag{1}$$

Here, $p'_{Du}(M_D;t_D) = \partial p_{Du}(M_D;t_D)/\partial t_D$.

I. *Production at a constant pressure*

If we now assume that the wellbore pressure, p_{wf}, is a constant, and if q_D given by

$$q_D(t_D) = \frac{q(t)\mu}{2\pi kh(p_i - p_{wf})}, \tag{2}$$

is the normalized rate for the constant-terminal-pressure case, then we obtain the important result (see van Everdingen and Hurst [2])

$$\overline{p}_{wDu}\overline{q}_D = \frac{1}{s^2}, \tag{3}$$

where the bars denote the Laplace transformation, and p_{wDu} is the wellbore response for the constant-terminal-rate case. This result will serve us in good stead when we consider layered geologic media. (3) is also important in that it permits us to simply write down the expressions for the production rate for the constant-terminal-pressure case from the constant-terminal-rate case. If the well is produced at a constant-terminal-pressure, the pressure distribution in the porous reservoir is given by

$$\overline{p}_D(M_D) = s\overline{q}_D\overline{p}_{Du}(M_D), \tag{4}$$

where $p_D = [p_i - p(M_D;t_D)]/(p_i - p_{wf})$, and $\overline{q}_D$ is given by (3). (4), of course, can be used for the general case when both the wellbore pressure and the production rate are functions of time and in this case $p_D(M_D;t_D)$ would have to be redefined.

II. *The wellbore-storage and skin effects*

Suppose V_w represents the volume of the wellbore, c is the compressibility of the fluid, ρ is the density of the fluid, and C represents the ability of the wellbore to store or unload fluid per unit change in pressure, that is $C = V_w c$. If we consider the wellbore to be the control volume, and if Δp_w is the change in pressure in the wellbore over a time δt, then from a mass balance, we have

$$\frac{k}{\mu}\int\int \rho\frac{\partial p}{\partial n}dS\delta t + (\rho C\Delta p)_w = (q\rho)_s\delta t. \tag{5}$$

Here, the subscripts s and w denote the surface and wellbore, respectively. The first term in (5) represents the influx from the porous medium to the wellbore, the second term reflects the ability of the wellbore to store fluids, and the third term reflects the production at the surface. Assuming density differences can be ignored, if q_{sf} represents influx into the wellbore, and $q = q_s$, then

$$q_{sf} + C\frac{d\Delta p_w}{dt} = q. \tag{6}$$

In terms of normalized pressure and normalized time, we have

$$\frac{q_{sf}}{q} + C_D\frac{dp_{wD}}{dt_D} = 1, \tag{7}$$

where

$$C_D = \frac{C}{2\pi\phi hc\ell^2}. \tag{8}$$

If times are large enough, we would not expect (5) to hold, and $q_{sf} = q$. In passing, we should note that the condition given in (7) is akin to the situation wherein we consider conduction of heat in a solid with its surface in contact with a well-stirred fluid (or a perfect conductor) so that the fluid temperature may be assumed to be constant throughout; see Carslaw and Jaeger [4].

Because of the manner in which wellbores are drilled, one has to contend, in practice, with the region of reduced permeability around the wellbore (skin effect). This region is treated in a manner similar to that of a poor conductor such as scale, grease or oxide that might exist on the surface of a conducting material in heat-conduction problems (see Carslaw and Jaeger [5]). Because of its size, one normally ignores the details of the properties of this region, and its effect is modelled via a normalized pressure, S. If this region is to be included, then the mathematical model satisfies the condition

$$p_{wD}(t_D) = p_D(M_{wD}+;t_D) + S\frac{q_{sf}}{q}. \tag{9}$$

Here, $p_D(M_{wD}+;t_D)$ represents the wellbore pressure responses derived in Chapters III and IV. If times are long enough, then $q_{sf}/q = 1$. Note that the pressure distribution in the reservoir is unaffected by the skin region.

Solutions for the well response, subject to the wellbore-storage boundary condition, can be obtained by Duhamel's formula. Combining (4), (7), and (9) after applying the Laplace transformation (assuming, of course, that (4) applies to the general case), we have the following result for the wellbore response:

$$\overline{p}_{wD} = \frac{\overline{p}_{wDu}}{1 + C_D s^2 \overline{p}_{wDu}}. \tag{10}$$

Here, $\overline{p}_{wDu}$ is the constant-terminal-rate solution and may include the existence of the skin effect. The important point here is that (10) applies to all the porous-media visualizations and wellbore conditions discussed in these notes (including those yet to be discussed).

5.2. Flow in naturally-fractured or fissured porous media

The classical treatment of flow in such a medium is given by Barenblatt, Zheltov, and Kochina [3]. In this scheme, the porous medium consists of a system of fissures and a system of porous, permeable blocks. The porous blocks are separated by the system of fissures, and the distribution of the fissures is such that direct diffusion between adjacent blocks is not possible. The fissures are assumed to have a negligible volume, and thus the blocks provide for the storage of mass. The flow capabilities of the fissures provide the flow path for diffusion to take place. Thus, in this scheme, at each point two fluid pressures exist—the pressure in the fissures, p_f, and the pressure in the blocks, p_m, and each pressure is an average over a neighborhood that contains a substantial number of blocks.

A slightly different scheme has been proposed by Kazemi [6] and further improved by de Swaan [7]. In this scheme (see Fig. 5.1), the fissure system and the matrix system consist of alternate layers of porous media with the storage capacity of the "fissure layers" being negligibly small when compared with the storage capacity of the "matrix layers". The key ingredient of this model is that horizontal flow in the fissure system produces a vertical gradient in the matrix system which in turns feeds the fissure system. Like the scheme proposed by Barenblatt *et al.* [3], the matrix blocks do not directly communicate with the wellbore. We shall now discuss the features of these models and demonstrate methods to incorporate them in the scheme we outlined in Chapters II and III.

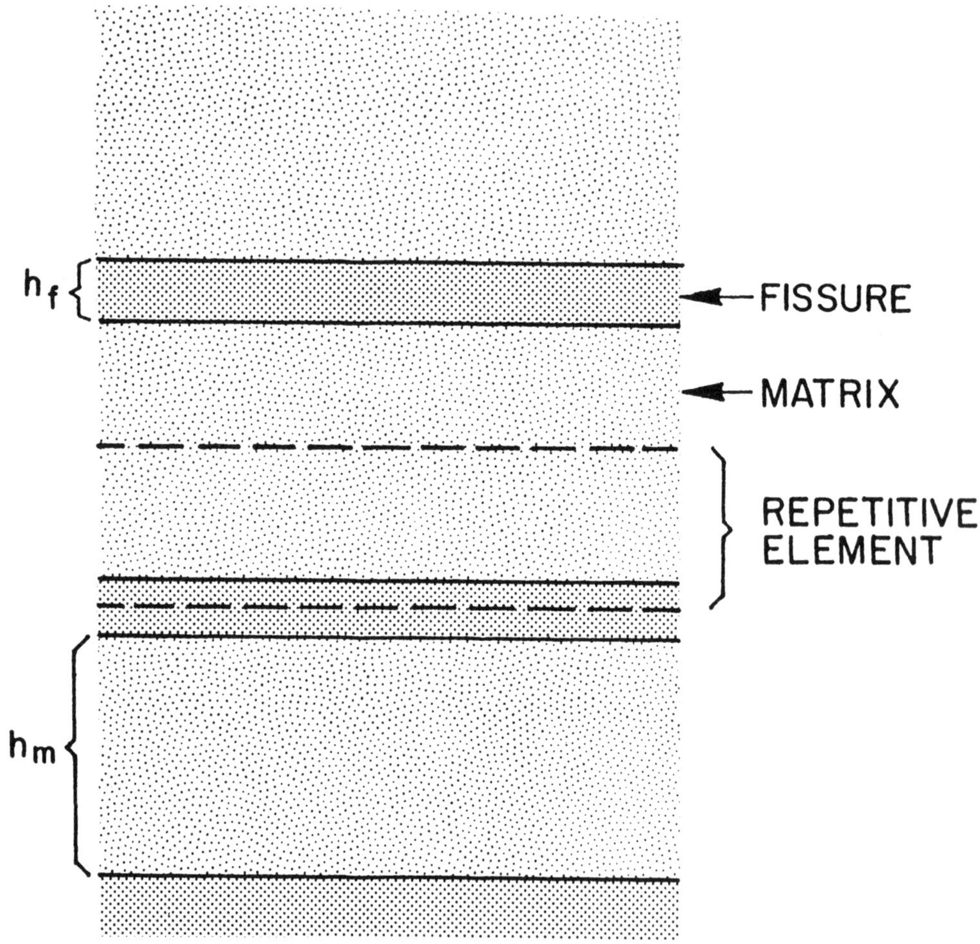

Figure 5.1. Fissured-Media Schematic of Kazemi (© courtesy SPE, from Ref. 8).

I. *The Barenblatt model*

The volume rate of the exchange of fluid between the blocks and fissures per unit volume of the porous medium is given by $k_m(p_m - p_f)/(\mu\epsilon)$, where $1/\epsilon$ is the characteristic of the medium that reflects the degree of fissuring or the surface area available for flow between the blocks and the fissures. For an isotropic system, the conservation of mass principle yields

$$\nabla \cdot (k_f \nabla \Delta p_f) = (V\phi c)_f \mu \frac{\partial \Delta p_f}{\partial t} + (V\phi c)_m \mu \frac{\partial \Delta p_m}{\partial t}. \tag{1}$$

Here, $\Delta p_j = p_i - p_j$ ($j = f, m$), and V_j is the ratio of the volume of System j to the total volume. Also,

$$\frac{k_m}{\mu\epsilon}(\Delta p_m - \Delta p_f) = -(V\phi c)_m \frac{\partial \Delta p_m}{\partial t}. \tag{2}$$

If we now define normalized time by

$$t_D = \frac{\eta t}{\ell^2}, \tag{3}$$

where

$$\eta = \frac{k_f}{[(V\phi c)_f + (V\phi c)_m]\mu}, \tag{4}$$

and the characteristic constants of the fissured porous medium λ and ω are given, respectively, by

$$\lambda = \frac{k_m}{\epsilon k_f}\ell^2 \tag{5}$$

and

$$\omega = \frac{(V\phi c)_f}{(V\phi c)_f + (V\phi c)_m}, \tag{6}$$

then we can write (1) and (2) as

$$\nabla_D^2 \Delta p_f = \omega \frac{\partial \Delta p_f}{\partial t_D} + (1-\omega)\frac{\partial \Delta p_m}{\partial t_D} \tag{7}$$

and

$$\lambda(\Delta p_m - \Delta p_f) = -(1-\omega)\frac{\partial \Delta p_m}{\partial t_D}, \tag{8}$$

respectively.

Applying the Laplace transformation to (7) and (8), we may write

$$\nabla_D^2 \overline{\Delta p}_f - s f(s) \overline{\Delta p}_f = 0, \tag{9}$$

where

$$f(s) = \frac{s\omega(1-\omega) + \lambda}{s(1-\omega) + \lambda}. \tag{10}$$

(9) is known as the fissured-medium equation.

If we replace u by $sf(s)$ in the kernals of the appropriate expressions, then we note that all the solutions that we have derived for the homogeneous reservoir case can be used to write expressions for the pressure distribution in the fissures of the system. Furthermore, from (10) we note that

$$f(s \to 0+) = 1, \tag{11}$$

and

$$f(s \to +\infty) = \omega. \tag{12}$$

A third approximation may also be used when $\lambda << (1-\omega)s$ and $\omega << \lambda/[(1-\omega)s]$. Under such circumstances we have (see Streltsova–Adams [9])

$$f(s) = \frac{\lambda}{(1-\omega)s}. \tag{13}$$

When (12) applies, support from the matrix system is negligible and the reservoir behaves as if it were a homogeneous system with a porosity-compressibility product equal to $(V\phi c)_f$. At late times, when $f(s \to 0+) = 1$, the system behaves as if it were a reservoir with an effective porosity–compressibility product equal to $(V\phi c)_f + (V\phi c)_m$. (13) applies at intermediate times and the wellbore pressure during this time period is approximately constant.

Based on this development, it is clear that by a simple change in the nomenclature, it is possible to derive a solution for the pressure distribution in the fissure system of a naturally-fractured reservoir by using the solutions that apply to the homogeneous reservoir. The pressure distribution in the matrix system can also be readily obtained.

II. *The Kazemi–de Swaan model*

Although first introduced by Kazemi [6], the impetus for much of the analytical development of this model is a result of de Swaan's [7] work. As noted earlier, in this scheme horizontal flow in the fissure system causes vertical flow in the matrix blocks. We consider one of the symmetrical elements in Figure 5.1. Application of the conservation of mass principle to a control volume in the fracture system yields

$$\nabla^2 \Delta p_f - \dot{q}_m = \frac{(V\phi c)_f \mu}{k_f} \frac{\partial \Delta p_f}{\partial t}. \tag{14}$$

Here, $\dot{q}_m$ is a source term that accounts for fluid influx from the blocks. The pressure distribution in a matrix block is given by

$$\frac{\partial^2 \Delta p_m}{\partial z^2} = \frac{(V\phi c)_m \mu}{k_m} \frac{\partial \Delta p_m}{\partial t}. \tag{15}$$

If we now scale z with respect to $h_m/2$, then (15) in terms of (3) may be written as

$$\frac{\partial^2 \Delta p_m}{\partial z_D^2} = (1-\omega)\frac{k_f h_m^2}{4k_m \ell^2} \frac{\partial \Delta p_m}{\partial t_D}. \tag{16}$$

The variable λ for this model is given by

$$\lambda = \frac{12 k_m \ell^2}{k_f h_m^2}, \tag{17}$$

and the variable ω is identical to that given in (6). Thus, (16) may be written as

$$\frac{\partial^2 \Delta p_m}{\partial z_D^2} = \frac{3(1-\omega)}{\lambda} \frac{\partial \Delta p_m}{\partial t_D}. \tag{18}$$

If we let $z_D = 0$ be the top of the repetitive element in Fig. 5.1 and $z_D = 1$ be the interface of the matrix block and the fissure, then using the no-flow boundary condition at $z_D = 0$ and the requirement that pressure be continuous at $z_D = 1$, we obtain

$$\overline{\Delta p_m} = \frac{ch\left(\sqrt{\tilde{\eta} s} z_D\right)}{ch\left(\sqrt{\tilde{\eta} s}\right)} \overline{\Delta p_f}, \tag{19}$$

where $\tilde{\eta} = 3(1-\omega)/\lambda$ and $ch(x)$ is the hyperbolic cosine of x. We can now show that

$$\bar{q}_m = \frac{1}{\ell^2} \sqrt{\frac{(1-\omega)\lambda s}{3}} th\left(\sqrt{\frac{3(1-\omega)s}{\lambda}}\right) \overline{\Delta p_f}. \tag{20}$$

Here, the symbol $th(x)$ represents the hyperbolic tangent of x. Using (20), we may write

$$\nabla_D^2 \overline{\Delta p_f} - s f(s) \overline{\Delta p_f} = 0, \tag{21}$$

where

$$f(s) = \omega + \sqrt{\frac{(1-\omega)\lambda}{3s}} th\left(\sqrt{\frac{3(1-\omega)s}{\lambda}}\right). \tag{22}$$

Again, this development emphasizes the fact that by a simple change in nomenclature the solutions for a homogeneous reservoir can be used for a fissured reservoir.

It is possible to obtain four approximations for $f(s)$. As $x \to 0+, th(x) \approx x$, and thus

$$f(s \to 0+) = 1. \tag{23}$$

Also, as $x \to +\infty$, $th(x) \approx 1$, and thus

$$f(s \to +\infty) = \omega. \tag{24}$$

If we assume that $\omega << 1$ and that $th(x) \approx 1$, we have (Bourdet and Gringarten [10], Dontsov and Boyrchuk [11], Streltsova [12], Serra, Reynolds, and Raghavan [13])

$$f(s) \approx \sqrt{\frac{(1-\omega)\lambda}{3s}}. \tag{25}$$

We may also use the approximation $th(x) \approx x - x^3/3$ in which case we have (Ozkan, Ohaeri, and Raghavan [8]),

$$f(s) = 1 - \frac{s(1-\omega)^2}{\lambda}. \tag{26}$$

5.3. Layered porous media

Here, we examine flow in a medium that may be considered to consist of distinct layers or strata that are separated by impervious layers. Communication between the layers is possible only because of the existence of wellbores. The properties of each of these layers are assumed to be distinct. Such a system is often referred to as a *commingled* or *no-crossflow* system. Our intent here is to obtain pressure distributions when the entire system is produced at a constant rate for combinations of wellbore configurations and outer-boundary conditions that may vary from layer to layer.

The algorithm to attain this objective, outlined by Spath, Ozkan, and Raghavan [14] is a rather simple one if we make note of the result obtained by van Everdingen and Hurst [2]; see 5.1(3). If we consider the constant-terminal-pressure case, then we note the simple fact that the layers are decoupled. Each layer produces fluid at a rate (against the common pressure) that is independent of the production rate of the other layers. Thus, if we let q_j represent the fluid withdrawal rate from Layer j, q be the total withdrawal rate, and n be the total number of layers, we have

$$q = \sum_{j=1}^{n} q_j. \tag{1}$$

If we now normalize the production rate q in terms of $\overline{kh}$ where $\overline{kh} = \sum_{j=1}^{n} k_j h_j$ and the production rate from Layer j by $k_j h_j$, then (1) may be written as

$$q_D = \sum_{j=1}^{n} \alpha_j q_{Dj}, \tag{2}$$

where $\alpha_j = k_j h_j / (\overline{kh})$ and the q_D's are normalized according to 5.1(2). Obtaining the wellbore response when the layered system produces at a constant-terminal-rate becomes a simple matter if we use 5.1(3). Thus, if $\overline{p}_{wD}$ is the Laplace transformation of the wellbore pressure, p_{wD}, then we have

$$\overline{p}_{wD} = \frac{1}{s^2 \sum_{j=1}^{n} \alpha_j \overline{q}_{Dj}}. \tag{3}$$

We may again use the relation given in 5.1(3) at the local level (at the level of the layers) and write

$$\overline{q}_{Dj} = \frac{1}{s^2 \overline{p}_{wDuj}} \tag{4}$$

and thus,

$$\overline{p}_{wD} = \frac{1}{\sum_{j=1}^{n} \alpha_j / \overline{p}_{wDuj}}. \tag{5}$$

(5) is a rather convenient result in that all the solutions we have derived thus far (including those for fissured or naturally-fractured porous media) can be used to obtain the wellbore response of a layered porous medium! Interestingly, no restrictions are placed on the wellbore or outer-boundary conditions of a given layer. For example, in a two-layer system we may assume that one layer is produced by an inclined well and the other layer by a horizontal well. Such schemes are not unusual. Wellbore-storage effects can be incorporated by considering 5.1(10). The skin effect in each layer should

be incorporated in the p_{wDuj} term. The only ramification we need to address is that we must express all solutions in terms of the same time scale which is a rather simple matter.

The expression for the production rate of each layer can also be obtained by Duhamel's [1] formula. Let the production rate of Layer j expressed as a fraction of the total production rate be $\widehat{q}_{Dj}$; that is, $\sum_{j=1}^{n} \widehat{q}_{Dj} = 1$. Thus, from Duhamel's formula we may write

$$\bar{\widehat{q}}_{Dj} = \alpha_j \frac{\bar{p}_{wD}}{s\bar{p}_{wDuj}}, \tag{6}$$

where $\bar{p}_{wD}$ is given in (5). Again, it is interesting that the layer rates can be obtained only by using wellbore information. Using (6), the pressure distribution in any layer may be computed by Duhamel's formula.

In some cases, because of conditions that have taken place over geologic time, it is possible for the initial pressures in the layers to be such that hydrostatic equilibrium does not exist. That is, for all practical purposes the initial pressures in the various layers are unequal at $t = 0$. The wellbore pressure in this case is given by

$$\bar{p}_{wD} = \frac{\sum_{j=1}^{n} \beta_j \bar{p}_{0Dj}}{\sum_{j=1}^{n} \beta_j} + \frac{1}{s^2 \sum_{j=1}^{n} \beta_j}, \tag{7}$$

where

$$\beta_j = \alpha_j \bar{q}_{Dj}, \tag{8}$$

and

$$p_{0Dj} = \frac{2\pi \bar{k} h}{q\mu}(p_r - p_{ji}). \tag{9}$$

Here, p_r is a reference pressure and p_{ji} is the initial pressure in Layer j. The sandface rate in Layer j is given by

$$\bar{\widehat{q}}_{Dj} = \frac{\alpha_j \left(\bar{p}_{wD} - \bar{p}_{0Dj}\right)}{s\bar{p}_{wDuj}}. \tag{10}$$

5.4. Wellbore hydraulics

Often it is advisable to couple flow in the wellbore with flow in the reservoir because fluids have to be produced at the surface. For illustrative purposes, we consider flow in a horizontal well of length, L_h. Because the volume of the wellbore is small, we consider steady flow within the wellbore. At the producing end ($x = 0$) the producing rate, q, is assumed to be a constant and the other end ($x = L_h$) is sealed. Fluid enters the wellbore at an unspecified rate along its length. Thus, if $q_{hc}(x)$ is the flow rate in the wellbore at any point x, and q_h is the flux, then

$$q_{hc}(x,t) = \int_x^{L_h} q_h(x',t)dx'. \tag{1}$$

Some insight into the variables that affect the performance of the system can be ascertained if we examine flow under steady conditions. If we consider steady flow of a Newtonian fluid in a pipe, we have

$$q^2 = \frac{\pi^2 r_w^5}{\rho f} \frac{dp_h}{dx}. \tag{2}$$

Here, ρ is the density of the fluid, f is the Fanning friction factor, r_w is the pipe radius and dp_h/dx is the pressure gradient. In terms of the Reynolds Number, Re, based on the pipe diameter, (2) may be written as

$$q = \frac{2\pi r_w^4}{\mu Re f} \frac{dp_h}{dx}. \tag{3}$$

If we assume that the porous medium is isotropic, then we have

$$\begin{aligned} q &= \frac{2\pi k}{\mu} r_w \int_0^{L_h} \left(\frac{\partial p}{\partial r}\right)_{r_w} dx' \\ &= \frac{2\pi k L_h r_w}{\mu} \left(\frac{\partial p}{\partial r}\right)_{r_w, x=\xi}, \end{aligned} \tag{4}$$

where $r^2 = x^2 + z^2$. From (3) and (4), we have

$$\frac{(\partial p/\partial r)_{r_w, x=\xi}}{dp_h/dx} = \frac{r_w^3}{k L_h Re\, f}. \tag{5}$$

Similarly, if we consider regions far from the horizontal well, we will find that

$$\frac{(\partial p/\partial r)_{x,y\to\infty}}{(\partial p_h/\partial r)_{r_w, x=\xi}} = \frac{r_w}{h}, \tag{6}$$

where h is the thickness of the porous medium. Thus, we expect a group of the form $r_w^4/(khL_h)$ to influence the pressure drop in the wellbore. Therefore, for convenience we will define dimensionless conductivity (that is akin to dimensionless fracture-conductivity) by the expression; see Ozkan, Sarica, Haciismoglu, and Raghavan [15]:

$$C_{hD} = \frac{\pi r_w^4}{8khL_h/2}, \tag{7}$$

where, in all of the following, $L_h/2$ is taken as the reference length. The mathematical development given below also confirms the arguments developed here on physical grounds. It is also clear that for laminar flow, C_{hD} will play an important role because the product $Re\ f =$ constant.

From (1), we may write

$$\frac{d^2 p_h}{dx^2} = \frac{\rho}{\pi^2 r_w^5} \left(q_{hc}^2 \frac{df}{dx} - 2f q_{hc} q_h \right). \tag{8}$$

Integrating (8) twice, we have

$$p_h(x;t) - p_{wf}(t) = \frac{\rho}{\pi^2 r_w^5}\left[f_t q^2 x + \int_{0+}^{x}\int_{0+}^{x'}\left(q_{hc}^2 \frac{df}{dx''} - 2 f q_{hc} q_h\right) dx'' dx'\right]. \qquad (9)$$

Here, f_t is the friction factor based on the total production rate q, and $p_{wf}(t)$ is the pressure at $x = 0$.

If we now use the definitions of normalized pressure, p_D, normalized time, t_D, and normalized distance, x_D, we may write (9) as

$$\begin{aligned} p_{wD}(t_D) - p_{hD}(x_D;t_D) &= \frac{Re_t f_t}{16}\frac{\pi}{C_{hD}}\Bigg[2x_D \\ &+ \int_{0+}^{x_D}\int_{0+}^{x'_D} \frac{Re}{Re_t}\frac{f}{f_t}\left(2\frac{Re}{f Re_t}\frac{df}{dx''_D} - 2q_{hD}\right) dx''_D dx'_D\Bigg]. \end{aligned} \qquad (10)$$

Here, the subscript t is used to denote conditions at $x_D = 0$, and $q_{hD} = q_h L_h / q$. Because

$$\frac{df}{dx_D} = -\frac{1}{2}\frac{df}{dRe} Re_t q_{hD} \qquad (11)$$

and if we let

$$D = Re^2 \frac{df}{dRe} + 2Re\ f, \qquad (12)$$

then

$$p_{wD}(t_D) - p_{hD}(x_D;t_D) = \frac{Re_t f_t}{8}\frac{\pi x_D}{C_{hD}} - \frac{\pi}{16 C_{hD}}\int_{0+}^{x_D}\int_{0+}^{x'_D} D q_{hD} dx''_D dx'_D. \qquad (13)$$

If flow is laminar everywhere, then $Re\ f = 16$; that is, $D = 16$, and (13) becomes

$$p_{wD}(t_D) - p_{hD}(x_D;t_D) = \frac{\pi}{C_{hD}}\left(2x_D - \int_{0+}^{x_D}\int_{0+}^{x''_D} q_{hD} dx''_D dx'_D\right). \qquad (14)$$

The numerical procedure to be followed when using (13) or (14) is similar to the one discussed in §3.1 V. We will now consider two cases to demonstrate the coupling of the wellbore and the porous medium. Let the horizontal well be parallel to the boundaries and located at an elevation, z_w. We consider flow in a slab reservoir with both boundaries impermeable.

I. *The long-time solution*

The long-time pressure-response for this system is given by

$$\begin{aligned} p_D(0 < x_D < 2, r_{wD}+; t_D) &= \frac{1}{2}\ln\frac{4t_D}{e^\gamma} \\ &+ \int_0^2 q_{hD}(x'_D)\left[\sigma(x_D, x'_D) + F(x_D, x'_D, z_{wD}, r_{wD}+, L_D)\right] dx'_D, \end{aligned} \qquad (15)$$

where

$$\sigma(x_D, x'_D) = -\frac{1}{2}\ln|x_D - x'_D| \tag{16}$$

and

$$F(x_D, x'_D, z_{wD}, r_{wD}, L_D) = \sum_{n=1}^{\infty} \cos n\pi(z_{wD} + r_{wD}) \cos n\pi z_{wD} K_0(n\pi L_D|x_D - x'_D|). \tag{17}$$

In writing (15), we have made use of the fact that

$$\int_0^2 q_{hD}(x'_D)dx'_D = 2. \tag{18}$$

For long enough times, rigorous calculations suggest that $q_{hD}(x_D;t_D)$ is independent of time. Thus, the right-hand side of (15) without the time-dependent term may be used for $p_{hD}(x_D;t_D)$ in (13) or in (14), and the resulting expression solved for q_{hD}. (The time-dependent terms cancel because the left-hand sides of (13) and (14) involve differences in pressures.)

II. *The large-s (short-time) approximation; laminar flow*

Let us now consider laminar flow in the wellbore and examine responses for times that are small enough. In this case q_{hD} is independent of x_D and is given by

$$q_{hD} = -2L_D\left(\tilde{r}_D\frac{\partial p_D}{\partial \tilde{r}_D}\right)_{\tilde{r}_D=\tilde{r}_{wD}}, \tag{19}$$

where $\tilde{r}_D^2 = y_D^2 + (z_D - z_{wD})^2/L_D^2$. Thus, the counterpart of (14) is given by

$$\frac{d^2\overline{p}_{hD}}{dx_D^2} - \frac{2\pi L_D\tilde{r}_{wD}\sqrt{s}}{C_{hD}}\frac{K_1(\sqrt{s}\tilde{r}_{wD})}{K_0(\sqrt{s}\tilde{r}_{wD})}\overline{p}_{hD} = 0. \tag{20}$$

Using the conditions

$$\left.\frac{d\overline{p}_{hD}}{dx_D}\right|_{x_D=2} = 0, \tag{21}$$

and

$$\left.\frac{d\overline{p}_{hD}}{dx_D}\right|_{x_D=0+} = -\frac{2\pi}{C_{hD}s}, \tag{22}$$

the pressure distribution in the wellbore is

$$\overline{p}_{hD}(x_D) = \frac{2\pi}{C_{hD}s\alpha}\frac{ch[\alpha(2-x_D)]}{sh(2\alpha)}, \tag{23}$$

where

$$\alpha^2 = \frac{2\pi L_D\tilde{r}_{wD}\sqrt{s}}{C_{hD}}\frac{K_1(\sqrt{s}\tilde{r}_{wD})}{K_0(\sqrt{s}\tilde{r}_{wD})}. \tag{24}$$

At the producing end ($x_D = 0$)

$$\overline{p}_{wD} = \frac{2\pi}{C_{hD}s\alpha\ th(2\alpha)}. \tag{25}$$

Approximate forms for p_{wD} can now be derived if s is assumed to be large enough. If $th(2\alpha) \approx 2\alpha$, then

$$\overline{p}_{wD} = \frac{1}{2L_{D}s}K_0(\sqrt{s}\widetilde{r}_{wD}), \tag{26}$$

and

$$p_{wD}(t_D) = -\frac{1}{4L_D}Ei\left(-\frac{\widetilde{r}^2_{wD}}{4t_D}\right). \tag{27}$$

Here, $-Ei(-\xi)$ is the exponential integral.

If α is such that $th(2\alpha) \approx 2\alpha - (2\alpha)^3/3$, then

$$\overline{p}_{wD} = \frac{1}{2L_{D}s}K_0\left(\sqrt{s}\widetilde{r}_{wD}\right) + \frac{4\pi}{3C_{hD}s} \tag{28}$$

and

$$p_{wD}(t_D) = -\frac{1}{4L_D}Ei\left(-\frac{\widetilde{r}^2_{wD}}{4t_D}\right) + \frac{4\pi}{3C_{hD}}. \tag{29}$$

References

1. Duhamel, J. M. C.: "Mémoire sur la méthod générale relative au mouvement de la chaleur dans les corps solides plongés dans les milieux dont la température varie avec le temps," *J. de Éc. Polyt.* (Paris) **14** (1833), 20–77.
2. van Everdingen, A. F., and Hurst, W.: "The Application of the LaPlace Transformation to Flow Problems in Reservoirs," *Transactions of AIME* (1949) **186**, 305–324.
3. Barenblatt, G. E., Zheltov, I. P., and Kochina, I. N: "Basic Concepts in the Theory of Homogeneous Liquids in Fissured Rocks," *Journal of Applied Mathematics and Mechanics* (1960), 1286–1303.
4. Carslaw, H. S., and Jaeger, J. C.: *Conduction of Heat in Solids*, Oxford University Press, Oxford (1959), 22.
5. Carslaw, H. S., and Jaeger, J. C.: *op. cit.*, 20.
6. Kazemi, H.: "Pressure Transient Analysis of Naturally Fractured Reservoirs," *Transactions of AIME* **256** (1960), 451–461.
7. de Swaan-O, A.: "Analytical Solutions for Determining Naturally Fractured Reservoir Properties by Well Testing," *Society of Petroleum Engineers Journal* (June 1976), 117–122.
8. Ozkan, E., Ohaeri, C. U., and Raghavan, R.: "Unsteady Flow to a Well Produced at Constant Pressure in a Fractured Reservoir," *SPE Formation Evaluation* (June 1987), 186–200.

9. Streltsova-Adams, T. D.: "Fluid Flow in Naturally Fractured Reservoirs," *Proceedings*, Second International Well Testing Symposium, Berkeley, California (October 25–27, 1978), 71–77.
10. Bourdet, D., and Gringarten, A. C.: "Determination of Fissure Volume and Block Size in Fractured Reservoirs," Paper SPE 9293 (1980).
11. Dontsov, K. M., and Boyrchuk, B. T.: "Effect of Characteristics of Fractured Media on Pressure Buildup Behavior," *Izvestia VUZ, Oil and Gas* (1971) N1, 42–46 (in Russian).
12. Streltsova, T. D.: "Well Pressure Behavior of a Naturally Fractured Reservoir," *Society of Petroleum Engineers Journal*, (October 1983), 769–780.
13. Serra, K., Reynolds, A. C., and Raghavan, R.: "New Pressure Transient Analysis Procedures for Naturally Fractured Reservoirs," *Journal of Petroleum Technology* (December 1983), 2271–2284.
14. Spath, J. B., Ozkan, E., and Raghavan, R.: "An Efficient Algorithm for Computation of Well Responses in Commingled Reservoirs," *SPE Formation Evaluation* (June 1994), 115–121.
15. Ozkan, E., Sarica, C., Haciislamoglu, M., Raghavan, R.: "Effect of Conductivity on Horizontal Well, Pressure Behavior", Paper SPE 24683 (1992).

Appendix–Fundamental Solution for Slab Reservoirs

Consider an infinite reservoir and let $O(x_{wD}, y_{wD}, z_{wD})$ denote the origin of a spherical coordinate system. If $P'(\rho'_D, \theta', \phi')$ is the location of the source, $P(\rho_D, \theta, \phi)$ is any point in the reservoir, and φ is the angle POP' where

$$\cos\varphi = \cos\theta\cos\theta' + \sin\theta\sin\theta'\cos(\phi - \phi'), \tag{1}$$

then the fundamental solution is given by

$$\overline{\gamma} = \frac{1}{4\pi}\sum_{m=0}^{\infty}\frac{(2m+1)}{\sqrt{\rho_D\rho'_D}}F\left(\sqrt{u}\rho'_D, \sqrt{u}\rho_D\right)P_m(\cos\varphi). \tag{2}$$

In (2), $P_m(x)$ represents the Legendre polynomial and we define

$$F(a,b) = \begin{cases} F_m(a,b); \; b < a \\ F_m(b,a); \; b > a, \end{cases} \tag{3}$$

where

$$F_m(a,b) = K_{m+\frac{1}{2}}(a)I_{m+\frac{1}{2}}(b). \tag{4}$$

Let us now consider the image system shown in Fig. A-1.

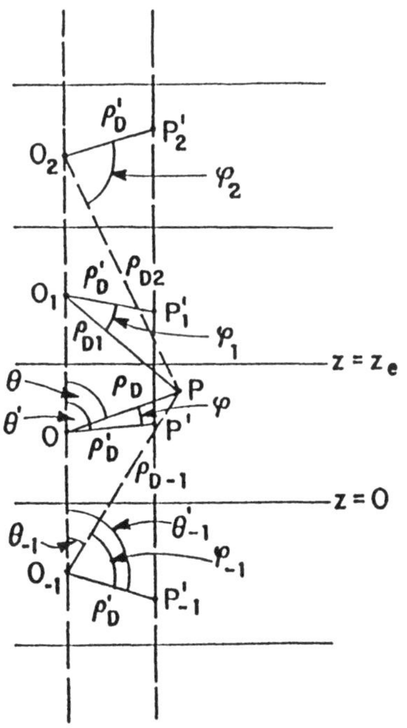

Figure A-1. Schematic of Image System

If we consider, temporarily, a new reference frame centered at $O_n(x_{wD}, y_{wD}, z_{wD} + nz_{eD})$, then the point P is defined by $P(\rho_{Dn}, \theta_n, \phi)$, the nth image of the source is located at $P'_n(\rho'_D, \theta'_n, \phi')$, and φ_n is the angle $P'_n O_n P$ where

$$\cos\varphi_n = \cos\theta_n \cos\theta'_n + \sin\theta_n \sin\theta'_n \cos(\phi - \phi'). \tag{5}$$

Therefore, the solution corresponding to the nth image of the source is given by

$$\overline{\gamma}_n = \frac{1}{4\pi} \sum_{m=0}^{\infty} \frac{(2m+1)}{\sqrt{\rho_{Dn}\rho'_D}} F\left(\sqrt{u}\rho'_D, \sqrt{u}\rho_{Dn}\right) P_m\left(\cos\varphi_n\right). \tag{6}$$

We may now relate ρ_{Dn}, θ_n, and θ'_n to ρ_D, θ, and θ'. Consider, for example, the image corresponding to $n = -1$ (see Fig. A-2).

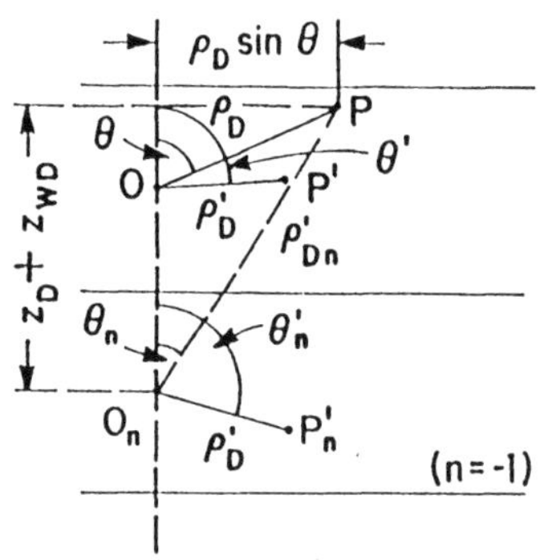

Figure A-2. Image for $n = -1$

We may write

$$\begin{aligned} \rho_{Dn}^2 &= (z_D - z_{wD} + 2z_{wD})^2 + \rho_D^2 \sin^2\theta \\ &= (\rho_D \cos\theta + 2z_{wD})^2 + \rho_D^2 \sin^2\theta, \end{aligned} \tag{7}$$

$$\cos\theta_n = \frac{z_D - z_{wD} + 2z_{wD}}{\rho_{Dn}} = \frac{\rho_D \cos\theta + 2z_{wD}}{\rho_{Dn}} \tag{8}$$

or

$$\theta_n = \arccos\frac{\rho_D \cos\theta + 2z_{wD}}{\rho_{Dn}}, \tag{9}$$

and

$$\theta'_n = \pi - \theta'. \tag{10}$$

By similar arguments, we may write

$$\rho_{Dn}^2 = \begin{cases} [\rho_D \cos\theta + 2z_{wD} - (n+1)z_{eD}]^2 + \rho_D^2 \sin^2\theta;\ n = \pm 1, \pm 3, \pm 5, \ldots \\ (\rho_D \cos\theta - nz_{eD})^2 + \rho_D^2 \sin^2\theta;\ n = 0, \pm 2, \pm 4, \ldots \end{cases} \tag{11}$$

$$\theta_n = \begin{cases} \arccos \dfrac{\rho_D \cos\theta + 2z_{wD} - (n+1)z_{eD}}{\rho_{Dn}}; \; n = \pm1, \pm3, \pm5, \ldots \\ \arccos \dfrac{\rho_D \cos\theta - nz_{eD}}{\rho_{Dn}}; \; n = 0, \pm2, \pm4, \ldots \end{cases} \tag{12}$$

$$\theta'_n = \begin{cases} \pi - \theta'; \; n = \pm1, \pm3, \pm5, \ldots \\ \theta'; \; n = 0, \pm2, \pm4, \ldots \end{cases} \tag{13}$$

Using (6), (11), (12), and (13), the fundamental solution for a slab reservoir with impermeable boundaries at $z = 0$ and $z = z_e$ may be written as

$$\overline{\gamma} = \frac{1}{4\pi\sqrt{\rho'_D}} \sum_{n=-\infty}^{+\infty} \sum_{m=0}^{\infty} (2m+1) \left[\frac{F_{mn1}}{\sqrt{\rho_{Dn1}}} P_m(\cos\varphi_{n1}) + \frac{F_{mn2}}{\sqrt{\rho_{Dn2}}} P_m(\cos\varphi_{n2}) \right], \tag{14}$$

where we have defined

$$F_{mk} = \begin{cases} K_{m+\frac{1}{2}}\left(\sqrt{u}\rho'_D\right) I_{m+\frac{1}{2}}\left(\sqrt{u}\rho_{Dk}\right); \; \rho_{Dk} < \rho'_D \\ K_{m+\frac{1}{2}}\left(\sqrt{u}\rho_{Dk}\right) I_{m+\frac{1}{2}}\left(\sqrt{u}\rho'_D\right); \; \rho_{Dk} > \rho'_D, \end{cases} \tag{15}$$

for $k = n1$ or $n2$, and

$$\rho^2_{Dn1} = (\rho_D \cos\theta + 2z_{wD} - 2nz_{eD})^2 + \rho^2_D \sin^2\theta, \tag{16}$$

$$\rho^2_{Dn2} = (\rho_D \cos\theta - 2nz_{eD})^2 + \rho^2_D \sin^2\theta, \tag{17}$$

$$\cos\varphi_k = \cos\theta_k \cos\theta'_k + \sin\theta_k \sin\theta'_k \cos(\phi - \phi'); \; k = n1 \text{ or } n2, \tag{18}$$

$$\theta_{n1} = \arccos \frac{\rho_D \cos\phi + 2z_{wD} - 2nz_{eD}}{\rho_{Dn1}}, \tag{19}$$

$$\theta_{n2} = \arccos \frac{\rho_D \cos\phi - 2nz_{eD}}{\rho_{Dn2}}, \tag{20}$$

$$\theta'_{n1} = \pi - \theta', \tag{21}$$

$$\theta'_{n2} = \theta'. \tag{22}$$

Similarly, if the boundaries at $z = 0$ and $z = z_e$ are at a constant pressure equal to the initial pressure, then we have

$$\overline{\gamma} = \frac{1}{4\pi\sqrt{\rho'_D}} \sum_{n=-\infty}^{+\infty} \sum_{m=0}^{\infty} (2m+1) \left[\frac{F_{mn2}}{\sqrt{\rho_{Dn2}}} P_m(\cos\varphi_{n2}) - \frac{F_{mn1}}{\sqrt{\rho_{Dn1}}} P_m(\cos\varphi_{n1}) \right]. \tag{23}$$

If the boundary at $z = 0$ is impervious and the boundary at $z = z_e$ is at a constant pressure equal to the initial pressure, then we obtain

$$\overline{\gamma} = \frac{1}{4\pi\sqrt{\rho'_D}} \sum_{n=-\infty}^{+\infty} \sum_{m=0}^{\infty} (2m+1)(-1)^n \left[\frac{F_{mn1}}{\sqrt{\rho_{Dn1}}} P_m(\cos\varphi_{n1}) + \frac{F_{mn2}}{\sqrt{\rho_{Dn2}}} P_m(\cos\varphi_{n2}) \right]. \tag{24}$$

Subject Index